CLIMATE CHANGE AND ITS CAUSES, EFFECTS AND PREDICTION

ECONOMIC COSTS OF INACTION ON CLIMATE CHANGE

ANALYSIS AND PERSPECTIVES

CLIMATE CHANGE AND ITS CAUSES, EFFECTS AND PREDICTION

Additional books in this series can be found on Nova's website under the Series tab.

Additional e-books in this series can be found on Nova's website under the e-book tab.

CLIMATE CHANGE AND ITS CAUSES, EFFECTS AND PREDICTION

ECONOMIC COSTS OF INACTION ON CLIMATE CHANGE

ANALYSIS AND PERSPECTIVES

CHERYL GRIFFIN
EDITOR

Copyright © 2014 by Nova Science Publishers, Inc.

All rights reserved. No part of this book may be reproduced, stored in a retrieval system or transmitted in any form or by any means: electronic, electrostatic, magnetic, tape, mechanical photocopying, recording or otherwise without the written permission of the Publisher.

For permission to use material from this book please contact us:
Telephone 631-231-7269; Fax 631-231-8175
Web Site: http://www.novapublishers.com

NOTICE TO THE READER

The Publisher has taken reasonable care in the preparation of this book, but makes no expressed or implied warranty of any kind and assumes no responsibility for any errors or omissions. No liability is assumed for incidental or consequential damages in connection with or arising out of information contained in this book. The Publisher shall not be liable for any special, consequential, or exemplary damages resulting, in whole or in part, from the readers' use of, or reliance upon, this material. Any parts of this book based on government reports are so indicated and copyright is claimed for those parts to the extent applicable to compilations of such works.

Independent verification should be sought for any data, advice or recommendations contained in this book. In addition, no responsibility is assumed by the publisher for any injury and/or damage to persons or property arising from any methods, products, instructions, ideas or otherwise contained in this publication.

This publication is designed to provide accurate and authoritative information with regard to the subject matter covered herein. It is sold with the clear understanding that the Publisher is not engaged in rendering legal or any other professional services. If legal or any other expert assistance is required, the services of a competent person should be sought. FROM A DECLARATION OF PARTICIPANTS JOINTLY ADOPTED BY A COMMITTEE OF THE AMERICAN BAR ASSOCIATION AND A COMMITTEE OF PUBLISHERS.

Additional color graphics may be available in the e-book version of this book.

Library of Congress Cataloging-in-Publication Data

ISBN: 978-1-61728-031-3

Published by Nova Science Publishers, Inc. † New York

Contents

Preface		vii
Chapter 1	The Cost of Delaying Action to Stem Climate Change *Council of Economic Advisers*	1
Chapter 2	Statement of Mindy Lubber, President, Ceres. Hearing on "The Costs of Inaction: The Economic and Budgetary Consequences of Climate Change" *Mindy Lubber*	39
Chapter 3	Budget Issues: Opportunities to Reduce Federal Fiscal Exposures Through Greater Resilience to Climate Change and Extreme Weather. Statement of Alfredo Gomez, Director, National Resources and Environment, U.S. Government Accountability Office. Hearing on "The Costs of Inaction: The Economic and Budgetary Consequences of Climate Change" *Alfredo Gomez*	47
Chapter 4	Testimony of Sherri W. Goodman, Executive Director, CNA Military Advisory Board. Hearing on "The Costs of Inaction: The Economic and Budgetary Consequences of Climate Change" *Sherri W. Goodman*	63

Chapter 5	Testimony of Bjørn Lomborg, Director, Copenhagen Consensus Center. Hearing on "The Costs of Inaction: The Economic and Budgetary Consequences of Climate Change" *Bjørn Lomborg*	75
Chapter 6	Testimony of W. David Montgomery, Senior Vice President, NERA Economic Consulting. Hearing on "The Costs of Inaction: The Economic and Budgetary Consequences of Climate Change" *W. David Montgomery*	97
Index		115

PREFACE

The changing climate and increasing atmospheric greenhouse gas (GHG) concentrations are projected to accelerate multiple threats, including more severe storms, droughts, and heat waves, further sea level rise, more frequent and severe storm surge damage, and acidification of the oceans. Beyond the sorts of gradual changes we have already experienced, global warming raises additional threats of large-scale changes, either changes to the global climate system, such as the disappearance of late-summer Arctic sea ice and the melting of large glacial ice sheets, or ecosystem impacts of climate change, such as critical endangerment or extinction of a large number of species. This book examines the cost of delaying policy actions to stem climate change.

Chapter 1 - The signs of climate change are all around us. The average temperature in the United States during the past decade was 0.8° Celsius (1.5° Fahrenheit) warmer than the 1901-1960 average, and the last decade was the warmest on record both in the United States and globally. Global sea levels are currently rising at approximately 1.25 inches per decade, and the rate of increase appears to be accelerating. Climate change is having different impacts across regions within the United States. In the West, heat waves have become more frequent and more intense, while heavy downpours are increasing throughout the lower 48 States and Alaska, especially in the Midwest and Northeast.1 The scientific consensus is that these changes, and many others, are largely consequences of anthropogenic emissions of greenhouse gases.2

The emission of greenhouse gases such as carbon dioxide (CO_2) harms others in a way that is not reflected in the price of carbon-based energy, that is, CO_2 emissions create a negative externality. Because the price of carbon-based energy does not reflect the full costs, or economic damages, of CO_2 emissions, market forces result in a level of CO_2 emissions that is too high.

Because of this market failure, public policies are needed to reduce CO2 emissions and thereby to limit the damage to economies and the natural world from further climate change.

There is a vigorous public debate over whether to act now to stem climate change or instead to delay implementing mitigation policies until a future date. This report examines the economic consequences of delaying implementing such policies and reaches two main conclusions, both of which point to the benefits of implementing mitigation policies now and to the net costs of delaying taking such actions.

Chapter 2 - Statement of Mindy Lubber, President, Ceres.

Chapter 3 - Statement of Alfredo Gomez, Director, National Resources and Environment, U.S. Government Accountability Office.

Chapter 4 - Testimony of Sherri W. Goodman, Executive Director, CNA Military Advisory Board.

Chapter 5 - Testimony of Bjørn Lomborg, Director, Copenhagen Consensus Center.

Chapter 6 - Testimony of W. David Montgomery, Senior Vice President, NERA Economic Consulting.

In: Economic Costs of Inaction on Climate Change ISBN: 978-1-61728-031-3
Editor: Cheryl Griffin   © 2014 Nova Science Publishers, Inc.

Chapter 1

THE COST OF DELAYING ACTION TO STEM CLIMATE CHANGE[*]

Council of Economic Advisers

EXECUTIVE SUMMARY

The signs of climate change are all around us. The average temperature in the United States during the past decade was 0.8° Celsius (1.5° Fahrenheit) warmer than the 1901-1960 average, and the last decade was the warmest on record both in the United States and globally. Global sea levels are currently rising at approximately 1.25 inches per decade, and the rate of increase appears to be accelerating. Climate change is having different impacts across regions within the United States. In the West, heat waves have become more frequent and more intense, while heavy downpours are increasing throughout the lower 48 States and Alaska, especially in the Midwest and Northeast.[1] The scientific consensus is that these changes, and many others, are largely consequences of anthropogenic emissions of greenhouse gases.[2]

The emission of greenhouse gases such as carbon dioxide (CO_2) harms others in a way that is not reflected in the price of carbon-based energy, that is, CO_2 emissions create a negative externality. Because the price of carbon-based energy does not reflect the full costs, or economic damages, of CO_2 emissions, market forces result in a level of CO_2 emissions that is too high. Because of this market failure, public policies are needed to

---

[*] This is an edited, reformatted and augmented version of a report issued July 2014.

reduce CO_2 emissions and thereby to limit the damage to economies and the natural world from further climate change.

There is a vigorous public debate over whether to act now to stem climate change or instead to delay implementing mitigation policies until a future date. This report examines the economic consequences of delaying implementing such policies and reaches two main conclusions, both of which point to the benefits of implementing mitigation policies now and to the net costs of delaying taking such actions.

First, although delaying action can reduce costs in the short run, on net, delaying action to limit the effects of climate change is costly. Because CO_2 accumulates in the atmosphere, delaying action increases CO_2 concentrations. Thus, if a policy delay leads to higher ultimate CO_2 concentrations, that delay produces persistent economic damages that arise from higher temperatures and higher CO_2 concentrations. Alternatively, if a delayed policy still aims to hit a given climate target, such as limiting CO_2 concentration to given level, then that delay means that the policy, when implemented, must be more stringent and thus more costly in subsequent years. In either case, delay is costly.

These costs will take the form of either greater damages from climate change or higher costs associated with implementing more rapid reductions in greenhouse gas emissions. In practice, delay could result in both types of costs. These costs can be large:

- Based on a leading aggregate damage estimate in the climate economics literature, a delay that results in warming of 3° Celsius above preindustrial levels, instead of 2°, could increase economic damages by approximately 0.9 percent of global output. To put this percentage in perspective, 0.9 percent of estimated 2014 U.S. Gross Domestic Product (GDP) is approximately $150 billion. The incremental cost of an additional degree of warming beyond 3° Celsius would be even greater. Moreover, these costs are not one-time, but are rather incurred year after year because of the permanent damage caused by increased climate change resulting from the delay.
- An analysis of research on the cost of delay for hitting a specified climate target (typically, a given concentration of greenhouse gases) suggests that net mitigation costs increase, on average, by approximately 40 percent for each decade of delay. These costs are higher for more aggressive climate goals: each year of delay means more CO2 emissions, so it becomes increasingly difficult, or even infeasible, to hit a climate target that is likely to yield only moderate temperature increases.

Second, climate policy can be thought of as "climate insurance" taken out against the most severe and irreversible potential consequences of climate change. Events such as the rapid melting of ice sheets and the consequent increase of global sea levels, or temperature increases on the higher end of the range of scientific uncertainty, could pose such severe economic consequences as reasonably to be thought of as climate catastrophes. Confronting the possibility of climate catastrophes means taking prudent steps now to reduce the future chances of the most severe consequences of climate change. The longer that action is postponed, the greater will be the concentration of CO_2 in the atmosphere and the greater is the risk. Just as businesses and individuals guard against severe financial risks by purchasing various forms of insurance, policymakers can take actions now that reduce the chances of triggering the most severe climate events. And, unlike conventional insurance policies, climate policy that serves as climate insurance is an investment that also leads to cleaner air, energy security, and benefits that are difficult to monetize like biological diversity.

I. INTRODUCTION

The changing climate and increasing atmospheric greenhouse gas (GHG) concentrations are projected to accelerate multiple threats, including more severe storms, droughts, and heat waves, further sea level rise, more frequent and severe storm surge damage, and acidification of the oceans (USGCRP 2014). Beyond the sorts of gradual changes we have already experienced, global warming raises additional threats of large-scale changes, either changes to the global climate system, such as the disappearance of late-summer Arctic sea ice and the melting of large glacial ice sheets, or ecosystem impacts of climate change, such as critical endangerment or extinction of a large number of species.

Emissions of GHGs such as carbon dioxide (CO_2) generate a cost that is borne by present and future generations, that is, by people other than those generating the emissions. These costs, or economic damages, include costs to health, costs from sea level rise, and damage from increasingly severe storms, droughts, and wildfires. These costs are not reflected in the price of those emissions. In economists' jargon, emitting CO_2 generates a negative externality and thus a market failure. Because the price of CO_2 emissions does not reflect its true costs, market forces alone are not able to solve the problem of climate change. As a result, without policy action, there will be more emissions and less investment in emissions-reducing technology than there would be if the price of emissions reflected their true costs.

This report examines the cost of delaying policy actions to stem climate change, and reaches two main conclusions. First, delaying action is costly. If a policy delay leads to higher ultimate CO_2 concentrations, then that delay produces persistent additional economic damages caused by higher temperatures, more acidic oceans, and other consequences of higher CO_2 concentrations. Moreover, if delay means that the policy, when implemented, must be more stringent to meet a given target, then it will be more costly.

Second, uncertainty about the most severe, irreversible consequences of climate change adds urgency to implementing climate policies *now* that reduce GHG emissions. In fact, climate policy can be seen as climate insurance taken out against the most damaging potential consequences of climate change—consequences so severe that these events are sometimes referred to as climate catastrophes. The possibility of climate catastrophes leads to taking prudent steps now to sharply reduce the chances that they occur.

The costs of inaction underscore the importance of taking meaningful steps today towards reducing carbon emissions. An example of such a step is the Environmental Protection Agency's (EPA) proposed rule (2014) to regulate carbon pollution from existing power plants. By adopting economically efficient mechanisms to reduce emissions over the coming years, this proposed rule would generate large positive net benefits, which EPA estimates to be in the range of $27 - 50 billion annually in 2020 and $49 - 84 billion in 2030. These benefits include benefits to health from reducing particulate emissions as well as benefits from reducing CO_2 emissions.

Delaying Climate Policies Increases Costs

Delaying climate policies avoids or reduces expenditures on new pollution control technologies in the near term. But this short-term advantage must be set against the disadvantages, which are the costs of delay. The costs of delay are driven by fundamental elements of climate science and economics. Because the lifetime of CO_2 in the atmosphere is very long, if a mitigation policy is delayed, it must take as its starting point a higher atmospheric concentration of CO_2. As a result, delayed mitigation can result in two types of cost, which we would experience in different proportions depending on subsequent policy choices.

First, if delay means an increase in the ultimate end-point concentration of CO2, then delay will result in additional warming and additional economic damages resulting from climate change. As is discussed in Section II, economists who have studied the costs of climate change find that temperature increases of 2° Celsius above preindustrial levels or less are likely to result in aggregate economic damages that are a small fraction of GDP. This small net effect masks important differences in which some regions could benefit somewhat from this warming while other regions could experience net costs. But global temperatures have *already* risen nearly 1° above preindustrial levels, and it will require concerted effort to hold temperature increases to within the narrow range consistent with small costs.[3] For temperature increases of 3° Celsius or more above preindustrial levels, the aggregate economic damages from climate change are expected to increase sharply.

Delay that causes a climate target to be missed creates large estimated economic damages. For example, a calculation in Section II of this report, based on a leading climate model (the DICE model as reported in Nordhaus 2013), shows that if a delay causes the mean global temperature increase to stabilize at 3° Celsius above preindustrial levels, instead of 2°, that delay will induce annual additional damages of approximately 0.9 percent of global output, as shown in Figure 1.[4] To put this percentage in perspective, 0.9 percent of estimated 2014 U.S. GDP is approximately $150 billion.[5] The next degree increase, from 3° to 4°, would incur greater *additional* annual costs of approximately 1.2 percent of global output. These costs are not one-time: they are incurred year after year because of the permanent damage caused by additional climate change resulting from the delay.

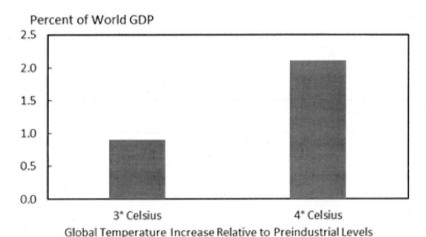

Source: Nordhaus (2013) and CEA calculations.

Figure 1. Economic Damage from Temperature Increase Beyond 2° Celsius.

The second type of cost of delay is the increased cost of reducing emissions more sharply if, instead, the delayed policy is to achieve the same climate target as the non-delayed policy. Taking meaningful steps now sends a signal to the market that reduces long-run costs of meeting the target. Part of this signal is that new carbon-intensive polluting facilities will be seen as bad investments; this reduces the amount of locked-in high-carbon infrastructure that is expensive to replace. Second, taking steps now to reduce CO_2 emissions signals the value of developing new low- and zero-emissions technologies, so additional steps towards a zero-carbon future can be taken as policy action incentivizes the development of new technologies. For both reasons, the least-cost mitigation path to achieve a given concentration target typically starts with a relatively low price of carbon to send these signals to the market, and subsequently increases as new low-carbon technology becomes available.[6]

The research discussed in Section II of this report shows that any short run gains from delay tend to be outweighed by the additional costs arising from the need to adopt a more abrupt and stringent policy later.[7] An analysis of the collective results from that research, described in more detail in Section II, suggests that the cost of hitting a specific climate target increases, on average, by approximately 40 percent for each decade of delay. These costs are higher for more aggressive climate goals: the longer the delay, the more difficult it becomes to hit a climate target. Furthermore, the research also finds that delay substantially decreases the chances that even concerted efforts in the future will hit the most aggressive climate targets.

Although global action is essential to meet climate targets, unilateral steps both encourage broader action and benefit the United States. Climate change is a global problem, and it will require strong international leadership to secure cooperation among both developed and developing countries to solve it. America must help forge a truly global solution to this global challenge by galvanizing international action to significantly reduce emissions. By taking credible steps toward mitigation, the United States will also reap the benefits of early action, such as investing in low-carbon infrastructure now that will reduce the costs of reaching climate targets in the future.

Climate Policy as Climate Insurance

Individuals and businesses routinely purchase insurance to guard against various forms of risk such as fire, theft, or other loss. This logic of self-protection also applies to climate change. Much is known about the basic science of climate change: there is a scientific consensus that, because of anthropogenic emissions of CO_2 and other GHGs, global temperatures are increasing, sea levels are rising, and the world's oceans are becoming more acidic. These and other climate changes are expected to be harmful, on balance, to the world's natural and economic systems. Nevertheless, uncertainty remains about the magnitude and timing of these and other aspects of climate change, even if we assume that future climate policies are known in advance. For example, the Working Group I contribution to the IPCC's Fifth Assessment Report (IPCC WG I AR5 2013) provides a likely range of 1.5° to 4.5° Celsius for the equilibrium climate sensitivity, which is the long-run increase in global mean surface temperature that is caused by a sustained doubling of atmospheric CO_2 concentrations. The upper end of that range would imply severe climate impacts under current emissions trajectories, and current scientific knowledge indicates that values in excess of this range are also possible.[8]

An additional, related source of climate uncertainty is the possibility of irreversible, large-scale changes that have wide-ranging and severe consequences. These are sometimes called abrupt changes because they could occur extremely rapidly as measured in geologic time, and are also sometimes called climate catastrophes. We are already witnessing one of these events—the rapid trend towards disappearance of late-summer Arctic sea ice. A recent study from the National Research Council (NRC 2013) found that this strong trend toward decreasing sea-ice cover could have large effects on a variety of components of the Arctic ecosystem and could potentially alter large-scale atmospheric circulation and its variability. The NRC also found that another large-scale change has been occurring, which is the critical endangerment or loss of a significant percentage of marine and terrestrial species. Other events judged by the NRC to be likely in the more distant future (after 2100) include, for example, the possible rapid melting of the Western Antarctic ice and Greenland ice sheets and the potential thawing of Arctic permafrost and the consequent release of the potent GHG methane, which would accelerate global warming. These and other potential large-scale changes are irreversible on relevant time scales—if an ice sheet melts, it cannot be reconstituted—and they could potentially have massive global consequences and costs. For many of these events, there is thought to be a "tipping point," for example a temperature threshold, beyond which the transition to the new state becomes inevitable, but the values or locations of these tipping points are typically unknown.

Section III of this report examines the implications of these possible climate-related catastrophes for climate policy. Research on the economic and policy implications of such threats is relatively recent. As detailed in Section III, a conclusion that clearly emerges from this young but active literature is that the threat of a climate catastrophe, potentially triggered by crossing an unknown tipping point, implies erring on the side of prudence today. Accordingly, in a phrase used by Weitzman (2009, 2012), Pindyck (2011), and others, climate policy can be thought of as "climate insurance." The logic here is that of risk management, in which one acts now to reduce the chances of worst-case outcomes in the future. Here, too, there is a cost to delay: the longer emission reductions are postponed, the greater are atmospheric concentrations of GHGs, and the greater is the risk arising from delay.

Other Costs of Delay and Benefits of Acting Now

An additional benefit of adopting meaningful mitigation policies now is that doing so sends a strong signal to the market to spur the investments that will reduce mitigation costs in the future. An argument sometimes made is that mitigation policies should be postponed until new low-carbon technologies become available. Indeed, ongoing technological progress has dramatically improved productivity and welfare in the United States because of vast inventions and process improvements in the private sector (see for example CEA 2014, Chapter 6). The private sector invests in research and development, and especially in process improvements, because those technological advances reap private rewards. But low-carbon technologies, and environmental technologies more generally, face a unique barrier: their benefits – the reduction in global impacts of climate change – accrue to everyone and not just to the developer or adopter of such technologies.[9] Thus private sector investment in low-carbon technologies requires confidence that those investments, if successful, will pay off, that is, the private sector needs to have confidence that there will be a market for low-carbon technologies now and in the future. Public policies that set out a clear and ongoing mitigation path provide that confidence. Simply waiting for a technological solution, but not providing any reason for the private sector to create that solution, is not an effective policy. Although public financing of basic research is warranted because many of the benefits of basic research cannot be privately appropriated, many of the productivity improvements and cost reductions seen in new technologies come from incremental advances and process improvements that only arise through private-sector experience producing the product and learning-by-doing. These advances are protected through the patent system and as trade secrets, but those advances will only transpire if it is clear that they will have current and future value. In other words, policy action induces technological change.[10] Although a full treatment of the literature on technological change is beyond the scope of this report, providing the private sector with the certainty needed to invest in low-carbon technologies and produce such technological change is a benefit of adopting meaningful mitigation policies now.

Finally, because this report examines the economic costs of delay, it focuses on actions or consequences that have a market price. But the total costs of climate change include much that does not trade in the market and to which it is difficult to assign a monetary value, such as the loss of habitat preservation, decreased value of ecosystem goods and services, and mass extinctions. Although some studies have attempted to quantify these costs, including all relevant climate impacts is infeasible. Accordingly, the monetized economic costs of delay analyzed in this report understate the true total cost of delaying action to mitigate climate change.

II. Costs from Delaying Policy Action

Delaying action on climate change can increase economic costs in two ways. First, if the delayed policy is no more stringent, it will miss the climate target of the original, non-delayed policy, resulting in atmospheric GHG concentrations that are permanently higher, thereby increasing the economic damages from climate change. Second, suppose a delayed policy alternatively strove to achieve the original climate target; if so, it would require a more stringent path to achieve that target. But this delayed, more stringent policy typically will result in additional mitigation costs by requiring more rapid adjustment later. In reality, delay might result in a mix of these two types of costs. The estimates of the costs of delay in this section draw on large bodies of research on these two types of costs. We first examine the economic damages from higher temperatures, then turn to the increased mitigation costs arising from delay.

Our focus here is on targets that limit GHG concentrations, both because this is what most of the "delay" literature considers and because concentration limits have been the focus of other assessments. These concentration targets are typically expressed as concentrations of CO_2- equivalent (CO_2e) GHGs, so they incorporate not just CO_2 concentrations but also methane and other GHGs. The CO_2e targets translate roughly into ranges of temperature changes as estimated by climate models and into the cumulative GHG emissions budgets discussed in some other climate literature. More stringent concentration targets decrease the odds that global average temperature exceeds 2°C above preindustrial levels by 2100. According to the IPCC WG III AR5 (2014), meeting a concentration target of 450 parts per million (ppm) CO2e makes it "likely" (probability between 66 and 100 percent) that the temperature increase will be at most 2°C, relative to preindustrial levels, whereas stabilizing at a concentration level of 550 ppm CO_2e makes it "more unlikely than likely" (less than a 50 percent probability) that the temperature increase by 2100 will be limited to 2°C (IPCC WG III AR5 2014).[11]

Increasing Damages if Delay Means Missing Climate Targets

If delay means that a climate target slips, then the ultimate GHG concentrations, temperatures, and other changes in global climate would be greater than without the delay.[12]

A growing body of work examines the costs that climate change imposes on specific aspects of economic activity. The IPCC WG II AR5 (2014) surveys this growing literature and summarizes the impacts of projected climate change by sector. Impacts include decreased agricultural production; coastal flooding, erosion, and submergence; increases in heat-related illness and other stresses due to extreme weather events; reduction in water availability and quality; displacement of people and increased risk of violent conflict; and species extinction and biodiversity loss. Although these impacts vary by region, and some impacts are not well-understood, evidence of these impacts has grown in recent years.[13]

A new class of empirical studies draw similar conclusions. Dell, Jones, and Olken (2013) review academic research that draws on historical variation in weather patterns to infer the effects of climate change on productivity, health, crime, political instability, and other social and economic outcomes. This approach complements physical science research by estimating the economic impacts of historical weather events that can be used to extrapolate to those expected in the future climate. The research finds evidence of economically meaningful impacts of climate change on a variety of outcomes. For example, when the temperature is greater than 100° Fahrenheit in the United States, labor supply in outdoor industries declines up to one hour per day relative to temperatures in the 76°-80° Fahrenheit range (Graff Zivin and Neidell 2014). Also in the United States, each additional day of extreme heat (exceeding 90° Fahrenheit) relative to a moderate day (50° to 59° Fahrenheit) increases the annual age-adjusted mortality rate by roughly 0.11 percent (Deschênes and Greenstone 2011).

These studies provide insights into the response of specific sectors or aspects of the economy to climate change. But because they focus on specific aspects of climate change, use different data sources, and use a variety of outcome measures, they do not provide direct estimates of the aggregate, or total, cost of climate change. Because estimating the total cost of climate change requires specifying future baseline economic and population trajectories, efforts to estimate the total cost of climate change typically rely on integrated assessment models (IAMs). IAMs are a class of economic and climate models that incorporate both climate and economic dynamics so that the climate responds to anthropogenic emissions and economic activity responds to the climate. In addition to projecting future climate variables and other economic variables, the IAMs estimate the total economic damages (and, in some cases, benefits) of climate change which includes impacts on agriculture, health, ecosystems services, productivity, heating and cooling demand, sea level rise, and adaptation.

Overall costs of climate change are substantial, according to IAMs. Nordhaus (2013) estimates global costs that increase with the rise in global average temperature, and Tol (2009, 2014) surveys various estimates. Two themes are common among these damage estimates. First, damage estimates remain uncertain, especially for large temperature increases. Second, the costs of climate change increase nonlinearly with the temperature change. Based on Nordhaus's (2013, Figure 22) net damage estimates, a 3° Celsius temperature increase above preindustrial levels, instead of 2°, results in additional damages of 0.9 percent of global output.[14] To put this percentage in perspective, 0.9 percent of estimated 2014 U.S. GDP is approximately $150 billion. The next degree increase, from 3° to 4°, would incur additional costs of 1.2 percent of global output. Moreover, these costs are not one-time, rather they recur year after year because of the permanent damage caused by increased climate change resulting from the delay. It should be stressed that these illustrative estimates are based on a single (albeit leading) model, and there is uncertainty associated with the aggregate monetized damage estimates from climate change; see for example the discussion in IPCC WG II AR5 (2014).

Increased Mitigation Costs from Delay

The second type of cost of delay arises if policy is delayed but still hits the climate target, for example stabilizing CO_2e concentrations at 550 ppm. Because a delay results in additional near-term accumulation of GHGs in the atmosphere, delay means that the policy, when implemented, must be more stringent to achieve the given long-term climate target. This additional stringency increases mitigation costs, relative to those that would be incurred under the least-cost path starting today.

This section reviews the recent literature on the additional mitigation costs of delay, under the assumption that both the original and delayed policy achieve a given climate target. We review 16 studies that compare 106 pairs of policy simulations based on integrated climate mitigation models (the studies are listed and briefly described in the Appendix). The simulations comprising each pair implement similar policies that lead to the same climate target (typically a concentration target but in some cases a temperature target) but differ in the timing of the policy implementation, nuanced in some cases by variation in when different countries adopt the policy. Because the climate target is the same for each scenario in the pair, the environmental and economic damages from climate change are approximately the same for each scenario. The additional cost of delaying implementation thus equals the difference in the mitigation costs in the two scenarios in each paired comparison. The studies reflect a broad array of climate targets, delayed timing scenarios, and modeling assumptions as discussed below. We focus on studies published in 2007 or later, including recent unpublished manuscripts.

In each case, a model computes the path of cost-effective mitigation policies, mitigation costs, and climate outcomes over time, constraining the emissions path so that the climate target is hit. Each path weighs technological progress in mitigation technology and other factors that encourage starting out slowly against the costs that arise if mitigation, delayed too long, must be undertaken rapidly. Because the models typically compute the policy in terms of a carbon price, the carbon price path computed by the model starts out relatively low and increases over the course of the policy. Thus a policy started today typically has a steadily increasing carbon price, whereas a delayed policy typically has a carbon price of zero until the start date, at which point it jumps to a higher initial level then increases more rapidly than the optimal immediate policy.

The higher carbon prices after a delay typically lead to higher total costs than a policy that would impose the carbon price today.[15]

The IPCC WG III AR5 (2014) includes an overview of the literature on the cost of delayed action on climate change. They cite simulation studies showing that delay is costly, both when all countries delay action and when there is partial delay, with some countries delaying acting alone until there is a more coordinated international effort. The present report expands on that overview by further analyzing the findings of the studies considered by the IPCC report as well as additional studies. Like the IPCC report, we find broad agreement across the scenario pairs examined that delayed policy action is more costly compared to immediate action conditional on a particular climate target. This finding is consistent across a range of climate targets, policy participants, and modeling assumptions. The vast majority of studies estimate that delayed action incurs greater mitigation costs compared to immediate action. Furthermore, some models used in the research predict that the most stringent climate targets are feasible only if immediate action is taken under full participation. One implication is that considering only comparisons with numerical cost estimates may understate the true costs of delay, as failing to reach a climate target means incurring the costs from the associated climate change.

The costs of delay in these studies depend on a number of factors, including the length of delay, the climate target, modeling assumptions, future baseline emissions, future mitigation technology, delay scenarios, the participants implementing the policy, and geographic location. More aggressive targets are more costly to achieve, and meeting them is predicted to be particularly costly, if not infeasible, if action is delayed. Similarly, international coordination in policy action reduces mitigation costs, and the cost of delay depends on which countries participate in the policy, as well as the length of delay.

The Role of Technological Progress in Cost Estimates

Assumptions about energy technology play an important role in estimating mitigation costs. For example, many models assume that carbon capture and storage (CCS) will enable point sources of emission to capture the bulk of carbon emissions and store them with minimal leakage into the atmosphere over a long period. Some comparisons also assume that CCS will combine with large-scale bio-energy ("bio-CCS"), effectively generating "negative emissions" since biological fuels extract atmospheric carbon during growth. Such technology could facilitate reaching a longterm atmospheric concentration target despite relatively modest near-term mitigation efforts. However, the IPCC warns that "There is only limited evidence on the potential for large-scale deployment of [bio-CCS], large-scale afforestation, and other [CO_2 removal] technologies and methods" (IPCC WG III AR5 2014).

> In addition, models must also specify the cost and timing of availability of such technology, potentially creating further variation in mitigation cost estimates.
>
> The potential importance of technology, especially bio-CCS, is manifested in differences across models. Clarke et al. (2009) present delay cost estimates for 10 models simulating a 550 ppm CO_2 equivalent target by 2100 allowing for overshoot. The three models that assume bio-CCS availability estimate global present values of the cost of delay ranging from $1.4 trillion to $4.7 trillion. Among the seven models without bio-CCS, four predict higher delay costs, one predicts that the concentration target was infeasible under a delay, and two predict lower delay costs. The importance of bio-CCS is even clearer with a more stringent target. For example, two of the three models with bio-CCS find that a 450 ppm CO_2 equivalent target is feasible under a delay scenario, while none of the seven models without bio-CCS find the stringent target to be feasible.
>
> The Department of Energy sponsors ongoing research on CCS for coal-fired power plants. As part of its nearly $6 billion commitment to clean coal technology, the Administration, partnered with industry, has already invested in four commercial-scale and 24 industrial-scale CCS projects that together will store more than 15 million metric tons of CO_2 per year.

An important determinant of costs is the role of technological progress and the availability of mitigation technologies (see the box). The models typically assume technological progress in mitigation technology, which means that the cost of reducing emissions declines over time as energy technologies improve. As a result, it is cost-effective to start with a relatively less stringent policy, then increase stringency over time, and the models typically build in this cost-effective tradeoff. However, most models still find that immediate initiation of a less stringent policy followed by increasing stringency incurs lower costs than delaying policy entirely and then increasing stringency more rapidly.

We begin by characterizing the primary findings in the literature broadly, discussing the estimates of delay costs and how the costs vary based on key parameters of the policy scenarios; additional details can be found in the Appendix. We then turn to a statistical analysis of all the available delay cost estimates that we could gather in a standardized form, that is, we conduct a meta-analysis of the literature on delay cost estimates.

Effect on Costs of Climate Targets, Length of Delay, and International Coordination

Climate Targets

Researchers estimate a range of climate and economic impacts from a given concentration of GHGs and find that delaying action is much costlier for more stringent targets. Two recent major modeling simulation projects conducted by the Energy Modeling Forum (Clarke et al. 2009) and by AMPERE (Riahi et al. 2014) consider the economic costs of delaying policies to reach a range of CO_2e concentration targets from 450 to 650 ppm in 2100. In the Energy Modeling Forum simulations in Clarke et al. (2009), the median additional cost (global present value) for a 20-year delay is estimated to be $0.7 trillion for 650 ppm CO_2e but a substantially greater $4.7 trillion for 550 ppm CO_2e. Many of the models in these studies suggest that delay causes a target of 450 ppm CO_2e to be much more costly to achieve, or possibly even infeasible.

Length of Delay

The longer the delay, the greater the cumulative emissions before action begins and the shorter the available time to meet a given target. Several recent studies examine the cost implications of delayed climate action and find that even a short delay can add substantial costs to meeting a stringent concentration target, or even make the target impossible to meet. For example, Luderer et al. (2012) find that delay from 2010 to 2020 to stabilize CO_2 concentration levels at 450 ppm by 2100 raises mitigation cost by 50 to 700 percent.[16] Furthermore, Luderer et al. find that delay until 2030 renders the 450 ppm target infeasible. Edmonds et al. (2008) find that additional mitigation costs of delay by newly developed and developing countries are substantial. In fact, they find that stabilizing CO_2 concentrations at 450 ppm even for a relatively short delay from 2012 to 2020 increases costs by 28 percent over the idealized case, and a delay to 2035 increased costs by more than 250 percent.

International Coordination

Meeting stringent climate targets with action from only one country or a small group of countries is difficult or impossible, making international coordination of policies essential. Recent research shows, however, that even if a delay in international mitigation efforts occurs, unilateral or fragmented action reduces the costs of delay: although immediate coordinated international action is the least costly approach, unilateral action is less costly than doing nothing.[17] More specifically, Jakob et al. (2012) consider a 10-year delay of mitigation efforts to reach a 450 ppm CO_2 target by 2100 and find that global mitigation costs increase by 43 to 700 percent if all countries begin mitigation efforts in 2020 rather than 2010. However, early action in 2010 by more developed countries reduces this increase to 29 to 300 percent. In a similar scenario, Luderer et al. (2012) find that costs increase by 50 to 700 percent with global delay from 2010 to 2020, however if the industrialized countries begin mitigation efforts unilaterally in 2010 (and are joined by all countries in 2020), the estimated cost increases range from zero to about 200 percent. Luderer et al. (2013) and Riahi et al. (2014) find that costs of delay are smaller when fewer countries delay mitigation efforts, or when short-term actions during the delay are more aggressive.

Jakob et al. (2012) find it is in the best interest of the European Union to begin climate action in 2010 rather than delaying action with all other countries until 2020. They also estimate that the cost increase to the United States from delaying climate action with all other countries until 2020 is from 28 to 225 percent, relative to acting early along with other industrialized economies.[18] McKibbin, Morris, and Wilcoxen (2014) consider the impact that a delay in imposing a unilateral price of carbon would have on economic outcomes in the United States including GDP, investment, consumption and employment. They find that although unilateral mitigation efforts do incur costs, delay is costlier.

Summary: Quantifying Patterns across the Studies

We now turn to a quantitative summary and assessment, or meta-analysis, of the studies discussed above.[19] The data set for this analysis consists of the results on all available numerical estimates of the average or total cost of delayed action from our literature search. Each estimate is a paired comparison of a delay scenario and its companion scenario without delay. To make results comparable across studies, we convert the delay cost estimates (presented in the original studies variously as present values of dollars, percent of consumption, or percent of GDP) to percent change in costs as a result of delay.[20] We capture variation across study and experimental designs using variables that encode the length of the delay in years; the target CO_2e concentration; whether only the relatively more-developed countries act immediately (partial delay); the discount rate used to calculate costs; and the model used for the simulation.[21] All comparisons consider policies and outcomes measured approximately through the end of the century. To reduce the effect of outliers, the primary regression analysis only uses results with less than a 400 percent increase in costs (alternative methods of handling the outliers are discussed below as sensitivity checks), and only includes paired comparisons for which both the primary and delayed policies are feasible (i.e. the model was able to solve for both cases).[22] The dataset contains a total of 106 observations (paired comparisons), with 58 included in the primary analysis. All observations in the data set are weighted equally.

Analysis of these data suggests two main conclusions, both consistent with findings from specific papers in the underlying literature. The first is that, looking across studies, costs increase with the length of the delay. Figure 2 shows the delay costs as a function of the delay time. Although there is considerable variability in costs for a given delay length because of variations across models and experiments, there is an overall pattern of costs increasing with delay.

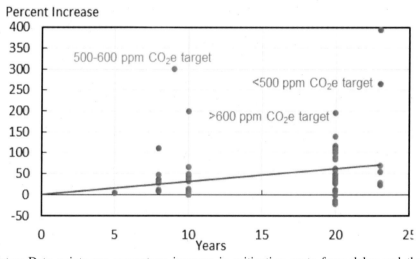

Notes: Data points are percentage increase in mitigation costs from delay and the associated length of delay for a given paired simulation. The scatterplot presents a total of 58 paired delay simulations. The solid line is the regression fit to these data.
Source. CEA calculations.

Figure 2. Additional Mitigation Costs of Delay by Length.

For example, of the 14 paired simulations with 10 years of delay (these are represented by the points in Figure 2 with 10 years of delay), the average delay cost is 39 percent. The regression line shown in Figure 2 estimates an average cost of delay per year using all 58 paired experiments under the assumption of a constant increasing delay cost per year (and, by definition, no cost if there is no delay), and this estimate is 37 percent per decade. This analysis ignores possible confounding factors, such as longer delays being associated with less stringent targets, and the multiple regression analysis presented below controls for such confounding factors.

The second conclusion is that the more ambitious the climate target, the greater are the costs of delay. This can be seen in Figure 3, in which the lowest (most stringent) concentration targets tend to have the highest cost estimates. In fact, close inspection of Figure 2 reveals a related pattern: the relationship between delay length and additional costs is steeper for the points representing CO_2e targets of 500 ppm or less than for those in the other two ranges. That is, costs of delay are particularly high for scenarios with the most stringent target and the longest delay lengths.

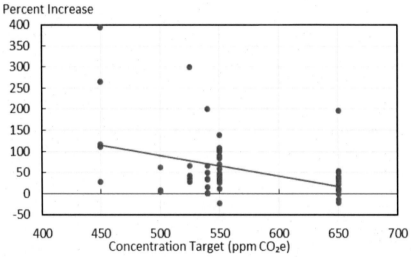

Notes: Data points are percentage increase in mitigation costs from delay and the associated CO_2 concentration target for a given paired simulation. The scatterplot presents a total of 58 paired delay simulations. The solid line is the regression line fit to these data.

Figure 3. Additional Mitigation Costs by CO_2 Concentration.

Table 1 presents the results of multiple regression analysis that summarizes how various factors affect predictions from the included studies, holding constant the other variables included in the regression. The dependent variable is the cost of delay, measured as the percentage increase relative to the comparable no-delay scenario, and the length of delay is measured in decades. Specifications (1) and (2) correspond to Figures 2 and 3, respectively. Each subsequent specification includes the length of the delay in years, an indicator variable for a partial delay scenario, and the target CO2e concentration. In addition to the coefficients shown, specification (4) includes model fixed effects, which control for systematic differences across models, and each specification other than column (1) includes an intercept.

The results in Table 1 quantify the two main findings mentioned above. The coefficients in column (3) indicate that, looking across these studies, a one decade increase in delay length is on average associated with a 41 percent increase in mitigation cost relative to the no-delay scenario. This regression does not control for possible differences in baseline costs across the different models, however, so column (4) reports a variant that includes an additional set of binary variables indicating the model used ("model fixed effects"). Including model fixed effects increases the delay cost to 56 percent per decade. When the cost of a delay is estimated separately for different concentration target bins (column (5)), delay is more costly the more ambitious is the concentration target. But even for the least ambitious target – a CO2e concentration exceeding 600 ppm – delay is estimated to increase costs by approximately 24 percent per decade. Because of the relatively small number of cases (58 paired comparisons), which are further reduced when delay is estimated within target bins, the standard errors are large, especially for the least ambitious scenarios, so for an overall estimate of the delay cost we do not differentiate between the different targets. While the regression in column (4) desirably controls for differences across models, other (unreported) specifications that handle the outliers in different ways and include other control variables give per-decade delay estimates both larger and smaller than the regression in column (3).[23] We therefore adopt the estimate in regression (3) of 41 percent per decade as the overall annual estimate of delay costs.

One caveat concerning this analysis is that it only considers cases in which model solutions exist. The omitted, infeasible cases tend to be ones with ambitious targets that cannot be met when there is long delay, given the model's technology assumptions. For this reason, omitting these cases arguably understates the costs of delay reported in Table 1.[24] Additionally, we note that estimates of the effect of a partial delay (when some developed nations act now and other nations delay action) are imprecisely estimated, perhaps reflecting the heterogeneity of partial delay scenarios examined in the studies.

Table 1. Increased Mitigation Costs Resulting from a Delay, Given a Specified Climate Target: Regression Results

	(1)	(2)	(3)	(4)	(5)
Delay (decades)	37.3***		41.1**	56.3***	
	(5.9)		(17.0)	(18.2)	
Delay (decades) x ppm $CO_2e \leq 500$					66.7**
					(27.1)
Delay (decades) x $500 <$ ppm $CO_2e \leq 600$					24.9
					(18.5)
Delay (decades) x ppm $CO_2e > 600$					24.1
					(33.9)
Partial delay			8.3	-20.0	14.8
			(26.0)	(27.8)	(25.7)
Target CO_2e concentration		-0.49***	-0.61***	-0.61***	-0.30
		(0.16)	(0.16)	(0.15)	(0.49)
Model fixed effects?	No	No	No	Yes	No
Observations	58	58	58	58	58
R-squared	0.41	0.15	0.24	0.53	0.30

Notes: The table presents ordinary least squares regression coefficients, with each column representing a different regression. For each, the dependent variable is the percent increase in cost from a scenario involving no delay to a scenario involving a delay. Each observation is a comparison of a pair of scenarios with the same climate target, for a total of 58 observations. The regressors represent some of the variables that characterize each paired comparison: the simulated delay, the delay interacted with the concentration target (binned), whether only some countries delayed (partial delay), and the target concentration. The appendix lists all studies from which the data were drawn. The specification in column (1) does not include a constant. Significant at the *10%**5%***1% significance level.
Source: CEA calculations on results from studies listed in appendix.

III. CLIMATE POLICY AS CLIMATE INSURANCE

As discussed in the 2013 NRC report, *Abrupt Impacts of Climate Change: Anticipating Surprises*, the Earth's climate history suggests the existence of "tipping points," that is, thresholds beyond which major changes occur that may be self-reinforcing and are likely to be irreversible over relevant time scales. Some of these changes, such as the rapid decline in late-summer Arctic sea ice, are already under way. Others represent potential events for which a tipping point likely exists, but cannot at the present be located. For example, there is new evidence that we might already have crossed a previously unrecognized tipping point concerning the destabilization of the West Antarctic Ice Sheet (Joughin, Smith, and Medley 2014 and Rignot et. al. 2014). A tipping point that is unknown, but thought unlikely to be reached in this century, is the release of methane from thawing Arctic permafrost, which could reinforce the greenhouse effect and spur additional warming and exacerbate climate change. Tipping points can also be crossed by slower climate changes that exceed a threshold at which there is a large-scale change in a biological system, such as the rapid extinction of species. Such impacts could pose such severe consequences for societies and economies that they are sometimes called potential climate catastrophes.

This section examines the implications of these potentially severe outcomes for climate policy, a topic that has been the focus of considerable recent research in the economics literature. The main conclusion emerging from this growing body of work is that the potential of these events to have large-scale impacts has important implications for climate policy. Because the probability of a climate catastrophe increases as GHG emissions rise, missing climate targets because of postponed policies increases risks. Uncertainty about the likelihood and consequences of potential climate catastrophes adds further urgency to implementing policies now to reduce GHG emissions.

Tail Risk Uncertainty and Possible Large-Scale Changes

Were some of these large-scale events to occur, they would have severe consequences and would effectively be irreversible. Because these events are thought to be relatively unlikely, at least in the near term – that is, they occur in the "tail" of the distribution – but would have severe consequences, they are sometimes referred to as "tail risk" events. Because these tail risk events are outside the range of modern human experience, uncertainty surrounds both the science of their dynamics and the economics of their consequences.

Because many of these events are triggered by warming, their likelihood depends in part on the equilibrium climate sensitivity. The IPCC WG I AR5 (2013) provides a likely range of 1.5° to 4.5° Celsius for the equilibrium climate sensitivity. However, considerably larger values cannot be ruled out and are more likely than lower values (i.e. the probability distribution is skewed towards higher values). Combinations of high climate sensitivity and high GHG emissions can result in extremely large end-of-century temperature changes. For example, the IPCC WG III AR5 (2014) cites a high-end projected warming of 7.8° Celsius by 2100, relative to 1900-1950.

A second way to express this risk is to focus on specific large-scale changes in Earth or biological systems that could be triggered and locked in by GHG concentrations rising beyond a certain point. At higher climate sensitivities, the larger temperature response to atmospheric GHG concentrations would make it even more likely that we would cross temperature-related tipping points in the climate system. The potential for additional releases of methane, a potent GHG, from thawing permafrost, thus creating a positive feedback to further increase temperatures, is an example of such a tail risk event. Higher carbon dioxide concentrations in the atmosphere, by increasing the acidity of the oceans, could also trigger and lock in permanent changes to ocean ecosystems, such as diminished coral reef-building, which decreases biodiversity supported on reefs and decreases the breakwater effects that protect shorelines. The probability of significant negative effects from ocean acidification can be increased by other stressors such as higher temperatures and overfishing.

The box summarizes some of these potential large-scale events, which are sometimes also referred to as "abrupt" because they occur in a very brief period of geological time. These events are sufficiently large-scale they have the potential for severely disrupting ecosystems and human societies, and thus are sometimes referred to as catastrophic outcomes.

Abrupt Impacts of Climate Change: Anticipating Surprises

The National Research Council's 2013 report, Abrupt Impacts of Climate Change: Anticipating Surprises, discusses a number of abrupt climate changes with potentially severe consequences. These events include:

- Late-summer Arctic sea ice disappearance: Strong trends of accelerating late-summer sea ice loss have been observed in the Arctic. The melting of Arctic sea ice comprises a positive feedback loop, as less ice means more sunlight will be absorbed into the dark ocean, causing further warming.
- Sea level rise (SLR) from destabilization of West Antarctic ice sheets (WAIS): The WAIS represents a potential SLR of 3-4 meters as well as coastal inundation and stronger storm surges. Much remains unknown of the physical processes at the ice-ocean frontier. However, two recent studies (Joughin, Smith, and Medley 2014, Rignot et. al. 2014) report evidence that irreversible WAIS destabilization has already started.

- Sea level rise from other ice sheets melting: Losing all other ice sheets, including Greenland, may cause SLR of up to 60 meters as well as coastal inundation and stronger storm surges. Melting of the Greenland ice sheet alone may induce SLR of 7m, but it is not expected to destabilize rapidly within this century.
- Disruption to Atlantic Meridional Overturning Circulation (AMOC): Potential disruptions to the AMOC may disrupt local marine ecosystems and shift tropical rain belts southward. Although current models do not indicate that an abrupt shift in the AMOC is likely within the century, the deep ocean remains understudied with respect to measures necessary for AMOC calculations.
- Decrease in ocean oxygen: As the solubility of gases decrease with rising temperature, a warming of the ocean will decrease the oxygen content in the surface ocean and expand existing Oxygen Minimum Zones. This will pose a threat to aerobic marine life as well as release nitrous oxide—a potent GHG—as a byproduct of microbial processes. The NRC study assesses a moderate likelihood of an abrupt increase in oxygen minimum zones in this century.
- Increasing release of carbon stores in soils and permafrost: Northern permafrost contains enough carbon to trigger a positive feedback response to warming temperatures. With an estimated stock of 1700-1800 Gt, the permafrost carbon stock could amplify considerably human-induced climate change. Small trends in soil carbon releases have been already observed.
- Increasing release of methane from ocean methane hydrates: This is a particularly potent long-term risk due to hydrate deposits through changes in ocean water temperature; the likely timescale for the physical processes involved spans centuries, however, and there is low risk this century.
- Rapid state changes in ecosystems, species range shifts, and species boundary changes: Research shows that climate change is an important component of abrupt ecosystem state-changes, with a prominent example being the Sahel region of Africa. Such state-changes from forests to savanna, from savanna to grassland, et cetera, will cause extensive habitat loss to animal species and threaten food and water supplies. The NRC study assesses moderate risk during this century and high risk afterwards.

> - Increases in extinctions of marine and terrestrial species: Abrupt climate impacts include extensive extinctions of marine and terrestrial species; examples such as the destruction of coral reef ecosystems are already underway. Numerous land mammal, bird, and amphibian species are expected to become extinct with a high probability within the next one or two centuries.

Implications of Tail Risk

An implication of the theory of decision-making under uncertainty is that the risks posed by irreversible catastrophic events can be substantial enough to influence or even dominate decisions.

Weitzman's Dismal Theorem

Over the past few years, economists have examined the implications of decision-making under uncertainty for climate change policy. In a particularly influential treatment, Weitzman (2009) proposes his so-called "Dismal Theorem," which provides a set of assumptions under which the current generation would be willing to bear very large (in fact, arbitrarily large) costs to avoid a future event with widespread, large-scale costs. The intuition behind Weitzman's mathematical result rests with the basic insight that because individuals are risk-averse, they prefer to buy health, home, and auto insurance than to take their chances of a major financial loss. Similarly, if major climate events have the potential to reduce aggregate consumption by a large amount, society will be better off if it can take out "climate insurance" by paying mitigation costs now that will reduce the odds of a large-scale—in Weitzman's (2009) word, catastrophic—drop in consumption later.[25]

Weitzman's (2009) dismal theorem has spurred a substantial amount of research on the economics of what this literature often refers to as climate catastrophes. A number of authors (e.g. Newbold and Daigneault 2009, Ackerman et al. 2010, Pindyck 2011, 2013, Nordhaus 2011, 2012, Litterman 2013, Millner 2013), including Weitzman (2011, 2014), stress that although the strong version of Weitzman's (2009) result—that society would be willing to pay an arbitrarily large amount to avoid future large-scale economic losses—depends on specific mathematical assumptions, the general principle of taking action to prevent such events does not. The basic insight is that, just as the sufficiently high threat of a fire justifies purchasing homeowners insurance, the threat of large-scale losses from climate change justifies purchasing "climate insurance" in the form of mitigation policies now (Pindyck 2011), and that taking actions today could help to avoid worst-case outcomes (Hwang, Tol, and Hofkes 2013). According to this line of thinking, the difficulty of assessing the probabilities of such large-scale losses or the location of tipping points does not change the basic conclusion that, because their potential costs are so overwhelming, the threat of very large losses due to climate change warrants implementing mitigation policies now.

Several recent studies have started down the road of quantifying the implications of the precautionary motive for climate policy. One approach is to build the effects of large-scale changes into IAMs, either by modeling the different risks explicitly or by simulation using heavy-tailed distributions for key parameters such as the equilibrium climate sensitivity or parameters of the economic damage function. Research along these lines includes Ackerman, Stanton, and Bueno (2013), Pycroft et al. (2011), Dietz (2011), Ceronsky et al. (2011), and Link and Tol (2011). Another approach is to focus on valuation of the extreme risks themselves outside an IAM, for example as examined by Pindyck (2012) and van der Ploeg and de Zeeuw (2013). Kopits, Marten, and Wolverton (2013) review some of the tail risk literature and literature on large-scale Earth system changes, and suggest steps forward for incorporating such events in IAMs, identifying ways in which the modeling could be improved even within current IAM frameworks and where additional work is needed. One of the challenges in assessing these large-scale events is that some of the most extreme events could occur in the distant future, and valuing consumption losses beyond this century raises additional uncertainty about intervening economic growth rates and questions about how to discount the distant future.[26] The literature is robust in showing that the potential for such events could have important climate policy implications, however, the scientific community has yet to derive robust quantitative policy recommendations based on a detailed analyses of the link between possible large-scale Earth system changes and their economic consequences.

Implications of Uncertainty about Tipping Points

Although research that embeds tipping points into climate models is young, one qualitative conclusion is that the prospect of a potential tipping point with unknown location enhances the precautionary motive for climate policy (Baranzini, Chesney, and Morisset 2003, Brozovic and Schlenker 2011, Cai, Judd, and Lontzek 2013, Lemoine and Traeger 2012, Barro 2013, van der Ploeg 2014). To develop the intuition, first suppose that the tipping point is a known temperature increase, say 3° Celsius above preindustrial levels, and that the economic consequences of crossing the tipping point are severe, and temporarily put aside other reasons for reducing carbon emissions. Under these assumptions climate policy would allow temperature to rise, stopping just short of the 3° increase. In contrast, now suppose that the tipping point is unknown and that its estimated mean is 3°, but that it could be less or more with equal probability. In this case, the policy that stops just short of 3° warming runs a large risk of crossing the true tipping point. Because that mistake would be very costly, the uncertainty about the tipping point generally leads to a policy that is more stringent today than it would be absent uncertainty. To the extent that delayed implementation means higher long-run CO2 concentrations, then the risks of hitting a tipping point increase with delay.

As a simplification, the above description assumes away other costs of climate change that increase smoothly with temperature, as well as the reality that important tipping points in biological systems could be crossed by small gradual changes in temperatures, so as to focus on the consequences of uncertainty about large-scale temperature changes. When the two sets of costs are combined, the presence of potential large-scale changes increases the benefits of mitigation policies, and the presence of uncertainty about tipping points that would produce abrupt changes increases those benefits further.[27] Cai, Judd, and Lontzek (2013) use a dynamic stochastic general equilibrium version of DICE model that is modified to include multiple tipping points with unknown (random) locations. To avoid the Weitzman "infinities" problem, they focus on tipping events with economic consequences that are large (5 or 10 percent of global GDP) but fall short of global economic collapses. They conclude that the possibility of future tipping points increases the optimal carbon price today: in their benchmark case, the optimal pre-tipping carbon price more than doubles, relative to having no tipping point dynamics. Similarly, Lemoine and Traeger (2012) embed unknown tipping points in the DICE model and estimate that the optimal carbon price increases by 45 percent as a result. In complementary work, Barro (2013) considers a simplified model in which the only benefits of reducing carbon emissions come from reducing the probability of potential climate catastrophes, and finds that this channel alone can justify investment in reducing GHG pollution of one percent of GDP or more, beyond what would normally occur in the market absent climate policy.

REFERENCES

Ackerman, Frank, Stephen J. DeCanio, Richard B. Howarth, and Kristen Sheeran. 2010. "The Need for a Fresh Approach to Climate Change Economics." In *Assessing the Benefits of Avoided Climate Change: Cost-Benefit Analysis and Beyond:* 159-181.

Ackerman, Frank, Elizabeth A. Stanton, and Ramón Bueno. 2013. "Epstein-Zin Utility in DICE: Is Risk Aversion Irrelevant to Climate Policy?" *Environmental Resource Economics* 56, 1: 73-84.

Barranzani, Andrea, Marc Chesney, and Jacques Morisset. 2003. "The Impact of Possible Climate Catastrophes on Global Warming Policy." *Energy Policy* 31, 8: 691-701.

Barro, Robert J. 2013. "Environmental Protection, Rare Disasters, and Discount Rates." *NBER Working Paper* 19258.

Blanford, Geoffrey J., Richard G. Richels, and Thomas F. Rutherford. 2009. "Feasible Climate Targets: The Roles of Economic Growth, Coalition Development and Expectations." *Energy Economics* 31, supplement 2: S82-S93.

Borenstein, Michael, Larry V. Hedges, Julian P.T. Higgins, and Hannah Rothstein. 2009. *Introduction to Meta-Analysis.* Chichester, U.K.: Wiley.

Bosetti, Valentina, Carlo Carraro, and Massimo Tavoni. 2009. "Climate Change Mitigation Strategies in Fast-Growing Countries: the Benefits of Early Action." *Energy Economics* 31, supplement 2: S14-S151.

Bosetti, Valentina, Carlo Carraro, Alessandra Sgobbi, and Massimo Tavoni. 2009. "Delayed Action and Uncertain Stabilisation Targets. How Much Will the Delay Cost?" *Climatic Change* 96, 3: 299-312.

Brozovic, N. and W. Schlenker. 2011. "Optimal Management of an Ecosystem with an Unknown Threshold." *Ecological Economics* 70, 4: 627-640.

Cai, Yonyang, Kenneth L. Judd, and Thomas S. Lontzek. 2013. "The Social Cost of Stochastic and Irreversible Climate Change." *NBER Working Paper* 18704.

Calvin, Katherine, James Edmonds, Ben Bond-Lamberty, Leon Clarke, Son H. Kim, Page Kyle, Steven J. Smith, Allison Thomson, and Marshall Wise. 2009a. "Limiting Climate Change to 450 ppm CO2 Equivalent in the 21st Century." *Energy Economics* 31, supplement 2: S107-S120.

Calvin, Katherine, Pralit Patel, Allen Fawcett, Leon Clarke, Karen Fisher-Vanden, Jae Edmonds, Son H. Kim, Ron Sands, and Marshall Wise. 2009b. "The Distribution and Magnitude of Emissions Mitigation Costs in Climate Stabilization under Less Than Perfect International Cooperation: SGM Results." *Energy Economics* 31, supplement 2: S187-S197.

Ceronsky, Megan, David Anthoff, Cameron Hepburn, and Richard S.J. Tol. 2011. "Checking the Price Tag on Catastrophe: The Social Cost of Carbon under Non-Linear Climate Response." *ESRI Working Paper* 392.

Clarke, Leon, Jae Edmonds, Volker Krey, Richard Richels, Steven Rose, and Massimo Tavoni. 2009. "International Climate Policy Architectures: Overview of the EMF 22 International Scenarios." *Energy Economics* 31, supplement 2: S64-S81.

Council of Economic Advisers. 2014. *Economic Report of the President, 2014.*

Dasgupta, Partha. 2008. "Discounting Climate Change." *Journal of Risk and Uncertainty* 37, 2/3: 141-169.

Dell, Melissa, Benjamin F. Jones, Benjamin A. Olken. 2013. "What Do We Learn from the Weather? The New Climate-Economy Literature." *Journal of Economic Literature*, forthcoming.

Deschênes, Olivier and Michael Greenstone. 2011. "Climate Change, Mortality, and Adaptation: Evidence from Annual Fluctuations in Weather in the US." *American Economic Journal: Applied Economics* 3, 4: 152-185.

Dietz, Simon. 2011. "High Impact, Low Probability? An Empirical Analysis of Risk in the Economics of Climate Change." *Climatic Change* 108, 3: 519-541.

Edmonds, Jae, Leon Clarke, John Lurz, and J. Macgregor Wise. 2008. "Stabilizing CO_2 Concentrations with Incomplete International Cooperation." *Climate Policy* 8, 4: 355- 376.

Graff Zivin, Joshua and Matthew Neidell. 2014. "Temperature and the Allocation of Time: Implications for Climate Change." *Journal of Labor Economics* 32, 1: 1-26.

Gurney, Andrew, Helal Ahammad, and Melanie Ford. 2009. "The Economics of Greenhouse Gas Mitigation: Insights from Illustrative Global Abatement Scenarios Modelling." *Energy Economics* 31, supplement 2: S174-S186.

Hwang, In Chang, Richard S.J. Tol, and Marjan W. Hofkes. 2013. "Tail-Effect and the Role of Greenhouse Gas Emissions Control." *University of Sussex Working Paper* Series 6613.

Intergovernmental Panel on Climate Change, Working Group I contribution to the Fifth Assessment Report (IPCC WG I AR5). 2013. *Climate Change 2013: The Physical Science Basis.*

Intergovernmental Panel on Climate Change, Working Group II contribution to the Fifth Assessment Report (IPCC WG II AR5). 2014. *Climate Change 2014: Impacts, Adaptation and Vulnerability.*

Intergovernmental Panel on Climate Change, Working Group III contribution to the Fifth Assessment Report (IPCC WG III AR5). 2014. *Climate Change 2014: Mitigation of Climate Change.*

Jaffe, Adam and Karen Palmer. 1997. "Environmental Regulation and Innovation: A Panel Data Study." *Review of Economics and Statistics* 79, 4: 610-619.

Jakob, Michael, Gunnar Luderer, Jan Steckel, Massimo Tavoni, and Stephanie Monjon. 2012. "Time to Act Now? Assessing the Costs of Delaying Climate Measures and Benefits of Early Action." *Climatic Change* 114, 1: 79-99.

Joughin, Ian, Benjamin E. Smith, and Brooke Medley. 2014. "Marine Ice Sheet Collapse Potentially Underway for the Thwaites Glacier Basin, West Antarctica." *Science* 344, 6185: 735-738.

Kopits, Elizabeth, Alex Marten, and Ann Wolverton. 2013. "Incorporating 'Catastrophic' Climate Change into Policy Analysis." *Climate Policy* ahead-of-print: 1-28.

Krey, Volker and Keywan Riahi. 2009. "Implications of Delayed Participation and Technology Failure for the Feasibility, Costs, and Likelihood of Staying Below Temperature Targets— Greenhouse Gas Mitigation Scenarios for the 21st Century." *Energy Economics* 31, supplement 2: S94-S106.

Lanjouw, Jean and Ashoka Mody. 1996. "Innovation and the International Diffusion of Environmentally Responsive Technology." *Research Policy* 25, 4: 549-571.

Lemoine, Derek and Christian Traeger. 2012. "Tipping Points and Ambiguity in the Economics of Climate Change." *NBER Working Paper* 18230.

Link, P. Michael and Richard S.J. Tol. 2011. "Estimation of the Economic Impact of Temperature Changes Induced by a Shutdown of the

Thermohaline Circulation: An Application of FUND." *Climactic Change* 104, 2: 287-304.

Litterman, Bob. 2013. "What is the Right Price for Carbon Emissions?" *Regulation* 36, 2: 38-51. Loulou, Richard, Maryse Labriet, and Amit Kanudia. 2009. "Deterministic and Stochastic Analysis of Alternative Climate Targets under Differentiated Cooperation Regimes." *Energy Economics* 31, supplement 2: S131-S143.

Luderer, Gunnar, Valentina Bosetti, Michael Jakob, Marian Leimbach, Jan Steckel, Henri Waisman, and Ottmar Edenhofer. 2012. "The Economics of Decarbonizing the Energy System – Results and Insights from the RECIPE Model Intercomparison." *Climatic Change* 114, 1: 9-37.

Luderer, Gunnar, Robert C. Pietzcker, Christoph Bertram, Elmar Kriegler, Malte Meinshausen, and Ottmar Edenhofer. 2013. "Economic Mitigation Challenges: How Further Delay Closes the Door for Achieving Climate Targets." *Environmental Research Letters* 8, 3.

McKibbin, Warwick J., Adele C. Morris, and Peter J. Wilcoxen. 2014. "The Economic Consequences of Delay in U.S. Climate Policy." Brookings: The Climate and Energy Economics Project.

Millner, Antony. 2013. "On Welfare Frameworks and Catastrophic Climate Risks." *Journal of Environmental Economics and Management* 65, 2: 310-325.

National Research Council. 2010. *Limiting the Magnitude of Future Climate Change*. Washington, D.C.: The National Academies Press

National Research Council. 2011. *Climate Stabilization Targets: Emissions, Concentrations, and Impacts over Decades to Millennia*. Washington D.C.: The National Academies Press. National Research Council. 2013. *Abrupt Impacts of Climate Change: Anticipating Surprises*. Washington D.C.: The National Academies Press.

Newbold, Stephen and Adam Daigneault. 2009. "Climate Response Uncertainty and the Benefits of Greenhouse Gas Emissions Reductions." *Environmental and Resource Economics* 44, 3: 351-377.

Nordhaus, William D. 2008. *A Question of Balance: Weighing the Options on Global Warming Policies*. New Haven: Yale University Press.

Nordhaus, William D. 2011. "The Economics of Tail Events with an Application to Climate Change." *Review of Environmental Economics and Policy* 5, 2: 240-257.

Nordhaus, William D. 2012. "Economic Policy in the Face of Severe Tail Events." *Journal of Public Economic Theory* 14, 2: 197-219.

Nordhaus, William D. 2013. *The Climate Casino: Risk, Uncertainty, and Economics for a Warming World.* New Haven: Yale University Press.
Pindyck, Robert S. 2011. "Fat Tails, Thin Tails, and Climate Change Policy." *Review of Environmental Economics and Policy* 5, 2: 258-274.
Pindyck, Robert S. 2012. "Uncertain Outcomes and Climate Change Policy." *Journal of Environmental Economics and Management* 63, 3: 289-303.
Pindyck, Robert S. 2013. "Climate Change Policy: What do the Models tell us?" *Journal of Economic Literature* 51, 3: 860-872.
Popp, David. 2003. "Pollution Control Innovations and the Clean Air Act of 1990." *Journal of Policy Analysis and Management* 22, 4: 641-660.
Popp, David. 2006. "International Innovation and Diffusion of Air Pollution Control Technologies: The effects of NOx and SO_2 Regulation in the U.S., Japan, and Germany." *Journal of Environmental Economics and Management* 51, 1: 46-71.
Popp, David, Richard G. Newell, and Adam B. Jaffe. 2010. "Energy, the Environment, and Technological Change." In *Handbook of the Economics of Innovation* 2: 873-937.
Pycroft, Jonathan, Lucia Vergano, Chris Hope, Daniele Paci, and Juano Carlos Ciscar. 2011. "A Tale of Tails: Uncertainty and the Social Cost of Carbon Dioxide." *Economics* 5, 22: 1-29.
Riahi, Keywan, Elmar Kriegler, Nils Johnson, Christoph Bertram, Michel den Elzen, Jiyong Eom, Michiel Schaeffer, Jae Edmonds, Morna Isaac, Volker Krey, Thomas Longden, Gunnar Luderer, Aurélie Méjean, David L. McCollum, Silvana Mimai, Hal Turton, Detlef P. van Vuuren, Kenichi Wada, Valentina Bosetti, Pantelis Caprosm, Patrick Criqui, Meriem Hamdi-Cherif, Mikiko Kainuma, and Ottmar Edenhofer. 2014. "Locked into Copenhagen Pledges—Implications of Short-Term Emission Targets for the Cost and Feasibility of Long-Term Climate Goals." *Technological Forecasting and Social Change.* In Press.
Richels, Richard G., Thomas F. Rutherford, Geoffrey J. Blanford, and Leon Clarke. 2007. "Managing the Transition to Climate Stabilization." *Climatic Policy* 7, 5: 409-428.
Rignot, Eric, Jeremie Mouginot, Mathieu Morlighem, Helene Seroussi, and Bernd Scheuchl. 2014. "Widespread, Rapid Grounding Line Retreat of

Pine Island, Thwaites, Smith, and Kohler Glaciers, West Antarctica, from 1992 to 2011." *Geophysical Research Letters* 41, 10: 3502-3509.

Roe, Gerard H. and Yoram Bauman. 2013. "Climate Sensitivity: Should the Climate Tail Wag the Policy Dog?" *Climatic Change* 117, 4: 647-662.

Russ, Peter and Tom van Ierland. 2009. "Insights on Different Participation Schemes to Meet Climate Goals." *Energy Economics* 31, supplement 2: S163-S173.

Tol, Richard S.J. 2009. "The Feasibility of Low Concentration Targets: An Application of FUND." *Energy Economics* 31, supplement 2: S121-S130.

Tol, Richard S.J. 2014. "Correction and Update: The Economic Effects of Climate Change." *Journal of Economic Perspectives* 28, 2: 221-226.

United States Environmental Protection Agency. 2014. "Carbon Pollution Emission Guidelines for Existing Stationary Sources: Electric Utility Generating Units." https://www.federalregister.gov/articles/2014/06/18/2014-13726/carbon-

(USGCRP) U.S. Global Change Research Program. 2014. *Climate Change Impacts in the United States: The Third National Climate Assessment.*

van der Ploeg, Frederick. 2014. "Abrupt Positive Feedback and the Social Cost of Carbon." *European Economic Review* 67: 28-41.

van der Ploeg, Frederick and Aart de Zeeuw. 2013. "Climate Policy and Catastrophic Change: Be Prepared and Avert Risk." *OxCarre Working Papers* 118, Oxford Centre for the Analysis of Resource Rich Economies, University of Oxford.

van Vliet, Jasper, Michael G.J. den Elzen, and Detlef P. van Vuuren. 2009. "Meeting Radiative Forcing Targets under Delayed Participation." *Energy Economics* 31, supplement 2: S152-S162.

Waldhoff, Stephanie A. and Allen A. Fawcett. 2011. "Can Developed Economies Combat Dangerous Anthropogenic Climate Change Without Near-Term Reductions from Developing Economies?" *Climatic Change* 107, 3/4: 635–641.

Waldhoff, Stephanie, Jeremy Martinich, Marcus Sarofim, Ben DeAngelo, James McFarland, Lesley Jantarasami, Kate Shouse, Allison Crimmins, Sara Ohrel, and Jia Li. 2014. "Overview of the Special Issue: A multi-model framework to achieve consistent evaluation of climate change impacts in the United States." *Climatic Change*, forthcoming.

Weitzman, Martin. 2009. "On Modeling and Interpreting the Economics of Catastrophic Climate Change." *The Review of Economics and Statistics* 91, 1: 1-19.

Weitzman, Martin. 2011. "Fat-Tailed Uncertainty in the Economics of Catastrophic Climate Change." *Review of Environmental Issues and Policy* 5, 2: 275-292.

Weitzman, Martin. 2012. "GHG Targets as Insurance against Catastrophic Climate Damages." *Journal of Public Economic Theory* 14, 2: 221-244.

Weitzman, Martin. 2013. "Tail-Hedge Discounting and the Social Cost of Carbon." *Journal of Economic Literature* 51, 3: 873-882.

Weitzman, Martin. 2014. "Fat Tails and the Social Cost of Carbon." *American Economic Review: Papers & Proceedings* 104, 5: 544-546.

APPENDIX: LITERATURE ON DELAY COSTS

This appendix lists the studies reviewed Section II and used in the meta-analysis, and briefly describes the scenarios they analyzed.

The EMF22 project engaged ten leading integrated assessment models to analyze the climate and economic consequences of delay scenarios. The EMF22 studies consist of Loulou, Labriet, and Kanudia (2009), Tol (2009), Gurney, Ahammad, and Ford (2009), van Vliet, den Elzen, and van Vuuren (2009), Blanford, Richels, and Rutherford (2009), Krey and Riahi (2009), Calvin et al. (2009a, 2009b), Russ and van Ierland (2009), and Bosetti, Carraro, and Tavoni (2009), with Clarke et al. (2009) providing an overview of the project.[28] Among other objectives, each study estimates the mitigation costs associated with five climate targets under both an immediate action scenario and a harmonized delay scenario. The targets are 450, 550, and 650 ppm CO_2e in 2100, and the models consider the first two targets alternatively allowing or prohibiting an overshoot before 2100.[29] In the delay scenario, only more developed countries (minus Russia) begin mitigation immediately in 2012 in a coordinated fashion (i.e., with the same carbon pricing), with some countries delaying action until 2030, and remaining countries delay action until 2050. These scenarios enable calculating the additional mitigation costs associated with delay for each concentration target.

The AMPERE project engaged nine modeling teams to analyze the climate and economic consequences of global emissions following the proposed policy stringency of the national pledges from the Copenhagen Accord and Cancún Agreements to 2030. (The AMPERE scenarios were not included in the meta-analysis in Section II because Riahi et al. (2014) did not provide sufficient information to calculate the percent increase in mitigation costs for each delay scenario.) One of the questions addressed by this project is the economic costs of delaying policies to reach CO2e concentration targets of 450 and 550 ppm in 2100 (Riahi et al. 2014). Eight models simulate pairs of policy scenarios reaching each target. One simulation in each pair assumes that all countries act immediately in a coordinated fashion (i.e., with the same carbon pricing), while the other simulation assumes that all countries follow the less stringent emissions commitments made during the Copenhagen Accord and Cancun Agreements until 2030, when coordinated international action begins.

The meta-analysis includes the following studies not associated with either AMPERE or EMF22: Jakob et al. (2012); Luderer et al. (2012, 2013); Edmonds et al. (2008); Richels et al. (2007), and Bosetti et al. (2009). Jakob et al. (2012) consider a 10-year delay of mitigation efforts to reach a 450 ppm CO2 target by 2100, including variations where more developed countries implement mitigation immediately. Luderer et al. (2012) consider a similar 10-year delay and the same 450 ppm CO2 target by 2100, with a scenario where Europe and all other industrialized countries begin mitigation efforts in 2010. Luderer et al. (2013) analyze a scenario where countries implement fragmented policies before coordinating efforts in 2015, 2020, or 2030 to meet a target of 2°C above preindustrial levels by 2100, allowing for overshooting. Edmonds et al. (2008) consider targets of 450, 550, and 660 ppm CO2, with newly developed and developing countries delaying climate action from a start date of 2012 to 2020, 2035 and 2050. Richels et al. (2007) estimate the additional cost of delay by newly developing countries until 2050 for a 450 and 550 ppm CO2 target. Finally, Bosetti et al. (2009) estimate the additional cost when all countries delay climate action for 20 years with a goal of reaching a 550 ppm and 650 ppm CO2e target by 2100.

End Notes

[1] For a fuller treatment of the current and projected consequences of climate change for U.S. regions and sectors, see the Third National Climate Assessment (United States Global Change Research Program (USGCRP) 2014).

[2] See for example the Summary for Policymakers in Working Group I contribution to the Intergovernmental Panel on Climate Change Fifth Assessment Report (IPCC WG I AR5 2013).

[3] The Working Group III contribution to the Intergovernmental Panel on Climate Change (IPCC) Fifth Assessment Report (IPCC WG III AR5 2014) does not analyze scenarios producing temperatures in 2100 less than 1.5 Celsius above preindustrial, because this is considered so difficult to achieve.

[4] Nordhaus (2013) stresses that these estimates "are subject to large uncertainties...because of the difficulty of estimating impacts in areas such as the value of lost species and damage to ecosystems." (pp. 139-140).

[5] These percentages apply to gross world output and the application of them to U.S. GDP is illustrative.

[6] The 2010 National Research Council, Limiting the Magnitude of Future Climate Change, also stressed the importance of acting now to implement mitigation policies as a way to reduce costs. The NRC emphasized the importance of technology development in holding down costs, including by providing clear signals to the private sector through predictable policies that support development of and investment in low-carbon technologies.

[7] The IPCC WG III AR5 (2014) includes an extensive discussion of mitigation, including sectoral detail, potential for technological progress, and the timing of mitigation policies.

[8] It is important to note that, as a global average, the equilibrium climate sensitivity masks the expectation that temperature change will be higher over land than the oceans, and that there will be substantial regional variations in temperature increases. The equilibrium climate sensitivity describes a long-term effect and is only one component of determining near term warming due to the buildup of GHGs in the atmosphere.

[9] Popp, Newell, and Jaffe (2010) provide a thorough review of the literature regarding technological change and the environment.

[10] For example, Popp (2003) provides empirical evidence that Title IV of the 1990 Clean Air Act Amendments (CAAA) led to innovations that reduced the cost of the environmental technologies that reduced SO2 emissions from coal-fired power plants. Other literature shows evidence linking environmental regulation more broadly to innovation (e.g., Popp 2006, Jaffe and Palmer 1997, Lanjouw and Mody 1996).

[11] IPCC WG III AR5 (2014, ch. 6) provides a further refinement of these probabilities, associating a concentration target of 450 ppm of CO2e with an approximate 70-85 percent probability of maintaining temperature change below 2°C, and a concentration level of 550 CO2e with an approximate 30-45 percent probability of maintaining temperature change below 2°C.

[12] For information on the impacts of climate change at various levels of warming see Climate Stabilization Targets: Emissions, Concentrations, and Impacts over Decades to Millennia (NRC 2011).

[13] The EPA's Climate Change Impacts and Risk Analysis project collects new research that estimates the potential damages of inaction and the benefits of GHG mitigation at national and regional scales for many important sectors, including human health, infrastructure, water resources, electricity demand and supply, ecosystems, agriculture, and forestry (Waldhoff et al. 2014).

[14] Some studies estimate that small temperature increases have a net economic benefit, for instance due to increased agricultural production in regions with colder climates. However, projected temperature increases even under immediate action fall in a range with a strong consensus that the costs of climate change exceed such benefits. The cost estimates presented here are net of any benefits expected to accrue.

[15] Some models explicitly identify the carbon price path that minimizes total social costs. These optimization models always find equal or greater costs for scenarios with a delay constraint. Other models forecast carbon prices that result in the climate target but do not demand that the path results in minimal cost. These latter models can predict that delay reduces costs, and a small number of comparisons we review report negative delay costs.

[16] We present a range of cost estimates which comes from the three IAMs – ReMIND-R, WITCH and IMACLIM-R – used by Luderer et al. (2012). These scenarios also allow temporary overshoot of the target.

[17] Waldhoff and Fawcett (2011) find that early mitigation action by industrialized economies significantly reduces the likelihood of large temperature changes in 2100 while also increasing the likelihood of lower temperature changes, relative to a no policy scenario.

[18] Note that the IMACLIM model finds that U.S. mitigation declines to the point in which they are slightly negative (i.e. net gains compared to business-as-usual).

[19] A study of the results of other studies is referred to as a meta-analysis, and there is a rich body of statistical tools for meta-analysis, see for example Borenstein et al. (2009).

[20] For example, if in some paired comparison delay increased mitigation costs from 0.20 percent of GDP to 0.30 percent of GDP, the cost increase would be 50 percent. Comparisons for which the studies provided insufficient information to calculate the percentage increase in costs (including all comparisons from Riahi et al. 2014) are excluded. Also excluded are comparisons that report only the market price of carbon emissions at the end of the simulation, which is not necessarily proportional to total mitigation costs.

[21] When measuring delay length for policies with multiple stages of implementation, we count the delay as ending at the start of any new participation in mitigation by any party after the start of the simulation. We also exclude scenarios with delays exceeding 30 years. When other climate targets were provided (e.g., CO2 concentration or global average temperature increase), the corresponding CO2e concentration levels are estimated using conversions from IPCC WG III AR5 (2014).

[22] In the event that a model estimates a cost for a first-best scenario but determines the corresponding delay scenario to be infeasible, the comparison is coded as having costs exceeding 400 percent. In addition, one comparison from Clarke et al. (2009) is excluded because a negative baseline cost precludes the calculation of a percent increase.

[23] The results in Table 1 are generally robust to using a variety of other specifications and regression methods, including: using the percent decrease from the delay case, instead of the percent increase from the no-delay case, as the dependent variable as an alternative way to handle outliers; using median regression, also as an alternative way to handle outliers; and including the discount factor as additional explanation of variation in the cost of delay, but this coefficient is never statistically significant. These regressions use linear compounding, not exponential, because the focus is on the per-decade delay cost not the annual delay cost. An alternative approach is to specify the dependent variable in logarithms (although this eliminates the negative estimates), and doing so yields generally similar results after compounding to those in Table 1.

[24] An alternative approach to omitting the infeasible-solution observations is to treat their values as censored at some level. Accordingly, the regressions in Table 1 were re-estimated using tobit regression, for which values exceeding 400 percent (including the non-solution cases) are treated as censored. As expected, the estimated costs of delay per year estimated by tobit regression exceed the ordinary least squares estimates. A linear probability model (not shown) indicates that scenarios with longer delay and more stringent targets are more likely to have delay cost increases exceeding 400 percent (including non-solution cases). The assumption of bio-CCS technology has no statistically significant correlation with delay cost increase in a censored regression but is associated with a significantly lower probability of delay cost increases exceeding 400 percent.

[25] This logic has its basis in expected utility theory. Because individuals are risk averse, each additional dollar of consumption provides less value, or utility, to individuals than the

previous dollar. To avoid this major loss, an individual will buy home insurance. That insurance is provided by the market because an insurance company can offer home insurance to many homeowners in different regions of the country, and through diversification the company will on average have many homeowners paying premiums and a few collecting insurance, so diversification allows the company to run a relatively low-risk business. But risks from severe climate change are not diversifiable because their enormous costs would impact the global economy. Consequently, as long as there is a non-negligible probability of a large drop in consumption, and therefore a very large drop in utility, arising from a large-scale loss in consumption, society today should be willing to pay a substantial amount if doing so would avoid that loss.

[26] For various perspectives on the challenges of evaluating long-term climate risks, see Dasgupta (2008), Barro (2013), Ackerman, Stanton, and Bueno (2013), Roe and Bauman (2013), and Weitzman (2013).

[27] Cai, Judd, and Lontzek (2013) provide a stark example of this dynamic. Their analysis, which is undertaken using a modified version of Nordhaus's (2008) DICE-2007 model, includes both the usual reasons for emissions mitigation (damages that increase smoothly with temperature) and the possibility of a tipping point at an uncertain future temperature which results in a jump in damages.

[28] Russ and van Ierland (2009) did not present estimates of total delay costs, so this paper is not included in the meta-analysis in Section II.

[29] We included three additional scenarios in van Vliet, den Elzen, and van Vuuren (2009) with alternate targets and models that were not reported in Clarke et al. (2009).

In: Economic Costs of Inaction on Climate Change ISBN: 978-1-61728-031-3
Editor: Cheryl Griffin   © 2014 Nova Science Publishers, Inc.

Chapter 2

STATEMENT OF MINDY LUBBER, PRESIDENT, CERES. HEARING ON "THE COSTS OF INACTION: THE ECONOMIC AND BUDGETARY CONSEQUENCES OF CLIMATE CHANGE"[*]

Chairman Murray, Ranking Member Sessions, members of the Committee, thank you for this opportunity to discuss with you the economic risks of climate change. My name is Mindy Lubber and I am the President of Ceres. Ceres is a national nonprofit organization mobilizing business and investor leadership on climate change and other global sustainability challenges. Ceres directs the Investor Network on Climate Risk, a network of 110 institutional investors with $13 trillion of collective assets focused on the risks of climate change. Ceres also coordinates BICEP - Business for Innovative Climate & Energy Policy - a network of 31 companies advocating for strong clean energy policies that includes major firms like Nike, Mars, Starbucks, Owens Corning, Jones Lang LaSalle, eBay, VF Corporation and General Mills.

The diversity of companies in BICEP represents the profound diversity of impacts that climate change is having on the U.S. economy – and the American taxpayers. For apparel giants VF Corp. and Nike, climate change poses risks to cotton and other commodities that are being affected by reduced water availability and drought. For Jones Lang LaSalle and Owens Corning,

---
[*] This is an edited, reformatted and augmented version of a statement presented July 29, 2014 before the Senate Budget Committee.

the climate change poses risks to buildings and their enormous use of electricity and growing vulnerability to coastal flooding and rising insurance costs. For General Mills and Starbucks, climate change poses risks to coffee, corn and other crucial crops that are experiencing more volatile growing conditions that oftentimes mean higher food prices. (We're seeing this right now, actually, with meat, fruit and vegetable prices all going up due to the prolonged drought in the West.)

Quite simply, climate change poses risks to every business sector and every American. The risks may vary, but they are being felt across our economy. That is why Ceres – and the companies and investors we work with – believe that the choice is NOT between protecting the climate and protecting the economy... We believe that without a stable climate, our economy cannot thrive.

The hundreds of companies and investors we work with all agree that climate change is a threat to their profitability and a threat to the global economy. These businesses are taking steps to prepare for the escalating impacts of climate change. They are pursuing sustainable technologies, such as using more renewable energy to slow climate change impacts and they are bringing their greenhouse gas emissions down. They are taking these steps because they believe the costs of not doing so are too great.

Ceres has done extensive research in the past decade on the many different ways that climate change is impacting our economy, hitting our wallets and creating bigger and bigger financial risks if actions are not taken.

Ceres has sponsored two reports, which are particularly relevant to today's discussion. In 2012 we published a report examining the growing costs and risks of extreme weather events; and last October, we published a report on the growing costs to taxpayers of inaction on climate change. I would like to include these two reports with my testimony for the record.

Ceres has identified five government disaster relief and recovery programs where the costs of inaction on climate change are most pronounced. They are federal disaster assistance appropriations, the National Flood Insurance Program, the Federal Crop Insurance Program, Wildfire Protection, and state-run insurance plans known as residual markets. Taxpayer bills and exposure for all of these programs are rising. Here are some numbers:

- First, with regard to federal disaster assistance appropriations, one conservative estimate puts the average bill that taxpayers can expect to pay at $20 billion a year. That's funding that goes to help communities respond to disasters such as hurricanes, thunderstorms and floods. But one should recognize that in any given year one catastrophic event alone could cost over $100 billion, causing that bill to the taxpayers to skyrocket. Hurricane Sandy, for example, cost Americans $60 billion in disaster relief costs.

- Second, our National Flood Insurance Program is currently in debt to the U.S. taxpayers for approximately $30 billion. In 2012, this vital program collected about $3.6 billion in premiums and paid out over $7.8 billion in Hurricane Sandy losses and other flood losses. While we'd like to think that Hurricane Sandy's devastating storm surge was an anomaly, it's not. Coastal flooding events are becoming more and more common, a result of rising sea levels and stronger storms, both of which are likely consequences of climate change. The average number of days per year that tidal waves have reached or surpassed flooding thresholds, that's a level when water begins collecting on surface streets – has more than tripled in many locations. Since 2001, water has hit these flooding thresholds an average of 20 days or more a year in several East Coast cities: Sandy Hook, N.J.; Atlantic City, N.J.; Annapolis, Md.; Washington, D.C.; Wilmington, N.C.; and Charleston, S.C.

The National Flood Insurance Program must increase premiums in these areas now prone to flood risks, or we can expect bigger program losses that American taxpayers will end up paying for.

- Third, the Federal Crop Insurance Program, a vital program that helps our farmers manage their risks, has seen insured crop losses spike from an average of $4.1 billion per year from 2001 through 2010, to a record-setting $10.8 billion in 2011. The devastating heat waves and drought in 2012 shattered even that record, when the program paid out $17.3 billion in crop losses

- Fourth, wildfire protection costs have tripled since the 1990s. Wildfire seasons are becoming longer and more severe. In the past 10 years federal government wildfire protection and suppression costs have averaged over $3 billion annually, compared to about $1 billion annually in the 1990s. FEMA's fire management assistance grants have more than tripled over the same period to an average of over $70 million annually. And state governments are spending up to another $2 billion annually on wildfire protection on top of the unknown amounts that local governments are spending.

- And, finally, state-run insurance plans known as residual insurance markets are facing dramatically larger loss exposure as private insurers pull out of states, especially coastal states, facing major climate risks. These state-run programs, backstopped by state taxpayers ultimately, have seen loss exposure grows from $54 billion in 1990 to $884.7 billion in 2011. I will come back to these state plans in more detail shortly.

If there is a moment historians will look back on as the moment when climate change truly hit home in America, it will almost surely be 2012. Add up all of the costs that I just enumerated– federal crop losses, flood losses, wildfire costs and disaster relief – extreme weather events cost Americans more than $300 per person in 2012 or $110 billion all together.

Yet, despite these rising losses in recent years, our federal disaster relief and recovery programs have been slow to recognize that worsening climate impacts will drive up future losses to unsustainable levels. Instead of encouraging behavior that reduces risks from extreme weather events, these programs encourage behavior that increases these risks – such as agricultural practices that increase vulnerability to drought and new development in hurricane- and wildfire-prone areas.

Citizens are not only paying for inaction on climate change through the increased costs of these federal programs. They are paying for them at the grocery store. Prolonged droughts in California, the Great Plains and the Southwest, have diminished the US cattle herd to its smallest size since 1951, causing beef prices to increase by 10% from a year ago. In decisions which have devastated many Texas communities, Cargill and other major livestock producers have been forced to shut down feedlot operations due to - as a Cargill spokesman put it – the "drought-depleted beef cattle supply."

Extreme weather is also contributing to prices for fresh fruits and eggs rising by five to six percent – twice the 2.8% rate of food price increases over the past 20 years.

According to the latest National Climate Assessment, released in May, the negative effects of climate change on agricultural production in the Midwest and Great Plains will far outweigh any positive effects. Corn production, the nation's biggest agriculture sector by far, is especially vulnerable to higher temperatures, changing rainfall patterns, soil erosion and water shortages that are widely predicted from climate change.

Competition for water, which is becoming ever more scarce in many US regions due to drier conditions, is especially pronounced in arid regions of the country, such as Southern California, the Southwest, and Texas. Ceres produced another study this year, "Hydraulic Fracturing and Water Stress:

Water Demand by the Numbers," that points out how hydraulic fracturing is increasing competitive pressures for water in some of the country's most water-stressed and drought-ridden regions. The report's review of hydraulic fracturing well data showed that 55 percent of the wells were in areas experiencing drought and 36 percent were in regions with significant groundwater depletion –key among those, California, which is in the midst of a horrific drought and Texas, which has the highest concentration of shale energy development and fracturing activity by far. Barring stiffer water-use regulations and improved on-the-ground practices, the industry's water needs in many regions are on a collision course with other water users, especially agriculture and municipal water use.

As climate change increases the risks of of extreme weather events, our federal and state disaster relief and insurance programs will become increasingly unsustainable. By one estimate, the net present value of the federal government's liability for unfunded disaster assistance over the next 75 years could actually be greater than the net present value of the unfunded liability for the Social Security program. [1]

Ceres has been working closely with state regulators in the insurance industry to set new standards and expectations that will enable insurers to plan for escalating climate risks while moving companies and individuals toward low-carbon activities. A growing number of companies in the sector recognize that climate change can have a devastating impact on their industry. They are the proverbial canary in the coal mine on climate change's impact on the economy.

Our report on the growing costs and risks of extreme weather states that, inevitably, as there is more weather damage, insurance companies, especially property & casualty firms, will charge more for their products. Ultimately this could lead to fewer people being able to afford insurance, as well as solvency problems for insurers themselves.

Insurance commissioners across the country are working with insurance companies to make sure that they are adequately addressing climate change in their risk profiles. From Washington to California to New York, insurance commissioners are mandating that major insurers disclose how they are managing the risks posed by climate change.

---

[1] J. David Cummins, Michael Suher and George Zanjani, "Federal Financial Exposure to Natural Catastrophe Risk," chapter in "Measuring and Managing Federal Financial Risk," National Bureau of Economic Research, edited by Deborah Lucas, February 2010.

Let me ask you, if you were told that there was a 98% chance that the boat you were about to board would sink, would you still climb aboard? The insurance industry has done this calculation with regard to climate change and they are not willing to take the risk that the 2% of scientists who are skeptical of climate change are right. They are preparing their industry for the long-term effects of severe weather due to climate change.

Unfortunately, from the taxpayers' vantage point, there's a down side to insurers' growing preparedness in risk-prone areas. Private insurers are especially leery of providing coverage in coastal areas vulnerable to more powerful hurricanes. In many regions, -- Florida, in particular – they've largely withdrawn from homeowners insurance markets because they were unable to charge substantially higher premiums. As a result, the states themselves are bearing the risks of providing homeowners insurance to millions of homeowners. Many coastal states are becoming so-called 'insurers of last resort.' What this means is that if a calamity strikes, state taxpayers will end up paying the bills.

According to the Government Accounting Office, from 1970 to 2010 state run insurance plans for those who cannot purchase insurance in the public market – so-called FAIR and Beach Plans – experienced explosive growth both in terms of policy count and exposure value. Total policies in force in the nation's FAIR, Beach and Windstorm Plans combined have more than tripled from 931,550 in 1990 to a record high 3.3 million in 2011. And as I said earlier, total loss exposure in these plans surged from $54.7 billion in 1990 to a record $884.7 billion in 2011—an increase of 1,517 percent – or 15 times.

At Ceres, while we clearly see the risks – both environmental and financial – from climate change, we also see opportunities in tackling the problem. Building a low-carbon economy will mean new job development and more investment in new businesses. We're also seeing compelling evidence of more and more American companies acting on climate change – and doing so affordably.

Last month, Ceres, Calvert and the World Wildlife Fund issued a "Power Forward 2.0" report showing how clean energy is becoming the mainstream for U.S. corporations. Sixty percent of the Fortune 100 companies have goals for renewable energy or greenhouse gas reductions. Through these initiatives, the 53 Fortune 100 companies reporting on climate and energy saving targets have collectively saved $1.1 billion annually and decreased their annual CO_2 emissions by the equivalent of retiring 15 coal-fired power plants.

We're also seeing impressive progress in the electric power sector, the largest source of CO_2 emissions in the country. Last week Ceres published a report benchmarking the country's largest 32 electric power companies on their energy efficiency and renewable energy programs. The report shows that electric utilities all over the country are delivering renewable energy and energy efficiency at scale. We found many strong performing utilities from all parts of the country such as Xcel Energy, a top ranking utility on renewable energy, which operates in Colorado, Minnesota, and Texas. And Pinnacle West in Arizona, which is achieving impressive energy savings for customers in a state that only recently began to set goals for these resources.

The report highlights recent studies showing that energy efficiency continues to rank as the lowest cost resource compared to all other electricity supply options. The report also cited a recent National Renewable Energy Lab study showing that renewable electricity has added only about 1 percent to electricity costs across the country. And renewable energy prices are continuing to drop at a rapid pace.

Tackling climate change offers one of the greatest economic opportunities of the 21st century—spurring innovation, creating good-paying jobs, and strengthening corporate bottom lines—all while protecting the economy from potentially catastrophic climate change impacts. More than 850 companies recognize this climate opportunity and are signatories to Ceres' Climate Declaration.

To truly seize this opportunity, we need to dramatically boost investments in clean energy and energy efficiency over the coming decades to cut carbon pollution and combat the worst effects of climate change. Globally, we need to achieve what we at Ceres call the Clean Trillion −$1 trillion in clean energy investing annually over the next 36 years.

Such investments globally are now at about a quarter-trillion dollars a year. So we have a long way to go. Meanwhile, the fossil fuel industry is spending over a half-trillion a year looking for new fossil fuel reserves that were we ever to burn them all would put us on a catastrophic path. The longer this paradigm continues, the longer we wait to tackle climate change with a vengeance, the more the costs—economic, human and environmental—will balloon.

I often talk about how when I had children, my whole outlook changed with regard to the environment. I really understood the importance of keeping the world safe for my kids and their kids.... Creating a world where they can live healthily whether they choose to live on Boston or Botswana. I feel the same way about the *economic* future of the world and of our country....and I believe they go hand in hand.

And I am sure that you members of the budget committee think about the importance of a strong economic future for our country every day. I would submit to you that a strong economic future depends on our country's response to the risk of climate change.

In: Economic Costs of Inaction on Climate Change   ISBN: 978-1-61728-031-3
Editor: Cheryl Griffin   © 2014 Nova Science Publishers, Inc.

Chapter 3

BUDGET ISSUES: OPPORTUNITIES TO REDUCE FEDERAL FISCAL EXPOSURES THROUGH GREATER RESILIENCE TO CLIMATE CHANGE AND EXTREME WEATHER. STATEMENT OF ALFREDO GOMEZ, DIRECTOR, NATIONAL RESOURCES AND ENVIRONMENT, U.S. GOVERNMENT ACCOUNTABILITY OFFICE. HEARING ON "THE COSTS OF INACTION: THE ECONOMIC AND BUDGETARY CONSEQUENCES OF CLIMATE CHANGE"[*]

WHY GAO DID THIS STUDY

Certain types of extreme weather events have become more frequent or intense according to the United States Global Change Research Program, including prolonged periods of heat, heavy downpours, and, in some regions, floods and droughts. While it is not possible to link any individual weather

---

[*] This is an edited, reformatted and augmented version of a statement presented July 29, 2014 before the Senate Budget Committee.

event to climate change, the impacts of these events affect many sectors of our economy, including the budgets of federal, state, and local governments.

GAO focuses particular attention on government operations it identifies as posing a "high risk" to the American taxpayer and, in February 2013, added to its High Risk List the area *Limiting the Federal Government's Fiscal Exposure by Better Managing Climate Change Risks*. GAO's past work has identified a variety of fiscal exposures—responsibilities, programs, and activities that may explicitly or implicitly expose the federal government to future spending.

This testimony is based on reports GAO issued from August 2007 to May 2014, and discusses (1) federal fiscal exposures resulting from climate-related and extreme weather impacts on critical infrastructure and federal lands, and (2) how improved federal technical assistance to all levels of government can help reduce climate-related fiscal exposures.

GAO is not making new recommendations but has made numerous recommendations in prior reports on this topic, which are in varying states of implementation by the Executive Office of the President and federal agencies.

WHAT GAO FOUND

Climate change and related extreme weather impacts on infrastructure and federal lands increase fiscal exposures that the federal budget does not fully reflect. Investing in resilience—actions to reduce potential future losses rather than waiting for an event to occur and paying for recovery afterward—can reduce the potential impacts of climate-related events. Implementing resilience measures creates additional up-front costs but could also confer benefits, such as a reduction in future damages from climate-related events. Key examples of vulnerable infrastructure and federal lands GAO has identified include:

- **Department of Defense (DOD) facilities.** DOD manages a global real-estate portfolio that includes over 555,000 facilities and 28 million acres of land with a replacement value DOD estimates at close to $850 billion. This infrastructure is vulnerable to the potential impacts of climate change and related extreme weather events. For example, in May 2014, GAO reported that a military base in the desert Southwest experienced a rain event in August 2013 in which about 1 year's worth of rain fell in 80 minutes. The flooding caused by the storm damaged more than 160 facilities, 8 roads, 1 bridge, and

11,000 linear feet of fencing, resulting in an estimated $64 million in damages.
- **Other large federal facilities.** The federal government owns and operates hundreds of thousands of other facilities that a changing climate could affect. For example, the National Aeronautics and Space Administration (NASA) manages more than 5,000 buildings and other structures. GAO reported in April 2013 that, in total, these NASA assets—many of which are in coastal areas vulnerable to storm surge and sea level rise—represent more than $32 billion in current replacement value.
- **Federal lands.** The federal government manages nearly 30 percent of the land in the United States—about 650 million acres of land—including 401 national park units and 155 national forests. GAO reported in May 2013 that these resources are vulnerable to changes in the climate, including the possibility of more frequent and severe droughts and wildfires. Appropriations for federal wildland fire management activities have tripled since 1999, averaging over $3 billion annually in recent years.

GAO has reported that improved climate-related technical assistance to all levels of government can help limit federal fiscal exposures. The federal government invests tens of billions of dollars annually in infrastructure projects that state and local governments prioritize, such as roads and bridges. Total public spending on transportation and water infrastructure exceeds $300 billion annually, with about 25 percent coming from the federal government and the rest from state and local governments. GAO's April 2013 report on infrastructure adaptation concluded that the federal government could help state and local efforts to increase their resilience by (1) improving access to and use of available climate-related information, (2) providing officials with improved access to technical assistance, and (3) helping officials consider climate change in their planning processes.

Chairman Murray, Ranking Member Sessions, and Members of the Committee:

I am pleased to be here today to discuss our work on reducing federal fiscal exposures posed by climate change and extreme weather events.[1] Climate change affects the American people in far-reaching ways, according to the National Research Council (NRC) and the United States Global Change Research Program's (USGCRP) May 2014 National Climate Assessment.[2] Certain types of extreme weather events with links to climate change have become more frequent or intense according to NRC and USGCRP, including prolonged periods of heat; heavy downpours; and, in some regions, floods and droughts. In addition, according to NRC and USGCRP, warming causes sea level to rise, sea ice to melt, and oceans to become more acidic as they absorb carbon dioxide. While it is not possible to link any individual weather event to climate change, these and other observed impacts of such events disrupt people's lives and affect many sectors of our economy, including the budgets of federal, state, and local governments.

Extreme weather events have cost the nation tens of billions of dollars in damages over the past decade. In 2012, for example, Superstorm Sandy alone caused tens of billions of dollars in damages to buildings, utilities, transportation systems, and other infrastructure. Heavy rainfall and snowfall events (which increase the risk of flooding) and heatwaves are generally becoming more frequent, consistent with theoretical expectations for a warmer and moister atmosphere due to changes in the climate, according to a February 2014 joint report by the U.S. National Academy of Sciences and the Royal Society in the United Kingdom.[3] The federal budget, however, generally does not account for disaster assistance provided in cases such as Superstorm Sandy—for which Congress provided about $60 billion in budget authority for such assistance—or the long-term impacts of climate change on existing federal infrastructure and programs.[4] Because of these significant financial risks and the nation's fiscal condition, in February 2013, we added *Limiting the Federal Government's Fiscal Exposure by Better Managing Climate Change Risks* to our list of high-risk areas.[5]

One way to reduce the potential impacts of climate change is to invest in enhancing resilience. The National Academies define resilience as the ability to prepare and plan for, absorb, recover from, and more successfully adapt to adverse events.[6] As we reported in April 2013, enhanced resilience results from actions to reduce potential future losses, rather than waiting for an event to occur and paying for recovery afterward.[7] Enhancing resilience has begun to receive more attention because greenhouse gases that are in the atmosphere could continue altering the climate system into the future, regardless of efforts to control emissions.

Implementing resilience measures creates additional up-front costs but could also confer benefits, such as a reduction in future damages from climate-related events. Federal efforts have begun to focus on enhancing resilience and providing information to state and local decision makers so they can make more informed decisions about fiscal exposure to potential climate-related events.[8] Decisions to adapt infrastructure to climate change can also depend on many other factors, such as the availability of substitutes or the remaining useful life of existing infrastructure.

My testimony today discusses (1) federal fiscal exposures resulting from climate-related and extreme weather impacts on critical infrastructure and federal lands, and (2) how improved federal technical assistance to all levels of government can help reduce climate-related fiscal exposures. My testimony is based on reports we issued from August 2007 to May 2014. Detailed information on our scope and methodology for our prior work can be found in those reports. The work this testimony is based on was conducted in accordance with generally accepted government auditing standards. Those standards require that we plan and perform the audit to obtain sufficient, appropriate evidence to provide a reasonable basis for our findings and conclusions based on our audit objectives. We believe that the evidence obtained provides a reasonable basis for our findings and conclusions based on our audit objectives.

CLIMATE-RELATED AND EXTREME WEATHER IMPACTS ON INFRASTRUCTURE AND FEDERAL LANDS INCREASE FEDERAL FISCAL EXPOSURES

As our past work has found, climate-related and extreme weather impacts on physical infrastructure such as buildings, roads, and bridges, as well as on federal lands, increase federal fiscal exposures. Infrastructure is typically designed to withstand and operate within historical climate patterns. However, according to NRC, as the climate changes, historical patterns do not provide reliable predictions of the future, in particular, those related to extreme weather events.[9] Thus, infrastructure designs may underestimate potential climate-related impacts over their design life, which can range up to 50 to 100 years. Federal agencies responsible for the long-term management of federal lands face similar impacts. Climate-related impacts can increase the operating and maintenance costs of infrastructure and federal lands or decrease the infrastructure's life span, leading to increased fiscal exposures for the federal government that are not fully reflected in the budget. Key examples from our recent work include (1) Department of Defense (DOD) facilities, (2) other large federal facilities such as National Aeronautics and Space Administration (NASA) centers, and (3) federal lands such as National Parks.

DOD Facilities

DOD manages a global real-estate portfolio that includes over 555,000 facilities and 28 million acres of land with a replacement value that DOD estimates at close to $850 billion. Within the United States, the department's extensive infrastructure of bases and training ranges— critical to maintaining military readiness—extends across the country, including Alaska and Hawaii. DOD incurs substantial costs for infrastructure, with a base budget for military construction and family housing totaling more than $9.8 billion in fiscal year 2014. As we reported in May 2014, this infrastructure is vulnerable to the potential impacts of climate change, including increased drought and more frequent and severe extreme weather events in certain locations.[10]

Source: Alaska Fire Service. | GAO-14-504T.
Note: Drought conditions contributed to a 2013 fire that limited the use of certain weapons systems and training activities.

Figure 1. Wildfire on a DOD Training Range in Alaska.

In its 2014 Quadrennial Defense Review, DOD stated that the impacts of climate change may increase the frequency, scale, and complexity of future missions, while undermining the capacity of domestic installations to support training activities. For example, in our May 2014 report on DOD infrastructure adaptation, we found that drought contributed to wildfires at an Army installation in Alaska that delayed certain units' training (see fig. 1).[11] Further, the fire limited the use of certain weapons systems in training and decreased the realism of the training.

Our May 2014 report also found that more frequent and more severe extreme weather events may result in increased fiscal exposure for DOD. Extreme precipitation events may lead to potential vulnerabilities such as increased maintenance costs for roads, utilities, and runways and increased flood-control measures. For example, we reported that in August 2013, a military base in the desert Southwest experienced an extreme rain event in which approximately 1 year's worth of rain fell in 80 minutes. According to Army officials and documents, the flooding caused by the storm damaged more than 160 facilities, 8 roads, 1 bridge, and 11,000 linear feet of fencing and resulted in an estimated $64 million in damage. Figure 2 shows flood damage to guard towers from this event.

Source: U.S. Army. | GAO-14-504T.
Note: Guard towers at an Army training area in the Southwestern United States (left); The same type of guard tower, toppled and severely damaged by flash flooding from an extreme precipitation event at this training area (right).

Figure 2. Army Training Area in Southwestern United States.

Other Large Federal Facilities

The federal government owns and operates hundreds of thousands of non-defense buildings and facilities that a changing climate could affect. For example, NASA's real property holdings include more than 5,000 buildings and other structures such as wind tunnels, laboratories, launch pads, and test stands. In total, these NASA assets—many of which are located in vulnerable coastal areas—represent more than $32 billion in current replacement value. Our April 2013 report on infrastructure adaptation showed the vulnerability of Johnson Space Center and its mission control center, often referred to as the nerve center for America's human space program.[12] As shown in figure 3, the center is located in Houston, Texas, near Galveston Bay and the Gulf of Mexico. Johnson Space Center's facilities—conservatively valued at $2.3 billion—are vulnerable to storm surge and sea level rise because of their location on the Gulf Coast.

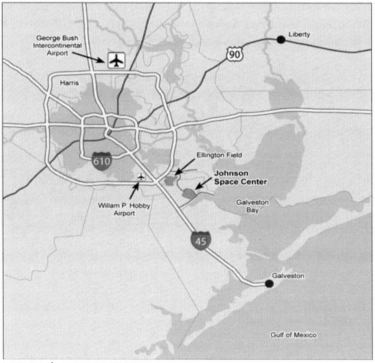

Source: NASA. | GAO-14-504T.

Figure 3. Location of Johnson Space Center.

Federal Lands

The federal government manages nearly 30 percent of the land in the United States for a variety of purposes, such as recreation, grazing, timber, and habitat for fish and wildlife. Specifically, federal agencies manage natural resources on about 650 million acres of land, including 401 national park units and 155 national forests. As we reported in May 2013, these resources are vulnerable to changes in the climate, including increases in air and water temperatures, wildfires, and drought; forests stressed by drought becoming more vulnerable to insect infestations; rising sea levels; and reduced snow cover and retreating glaciers.[13] In addition, various species are expected to be at risk of becoming extinct due to the loss of habitat critical to their survival. Many of these changes have already been observed on federally managed lands and waters and are expected to continue, and one of the areas where the federal government's fiscal exposure is expected to increase is in its role as the manager of large amounts of land and other natural resources. According to USGCRP's May 2014 National Climate Assessment, hotter and drier weather and earlier snowmelt mean that wildfires in the West start earlier in the spring, last later into the fall, and burn more acres.[14] Appropriations for the federal government's wildland fire management activities have tripled, averaging over $3 billion annually in recent years, up from about $1 billion in fiscal year 1999.[15]

IMPROVED CLIMATE-RELATED TECHNICAL ASSISTANCE TO ALL LEVELS OF GOVERNMENT CAN HELP LIMIT FEDERAL FISCAL EXPOSURES

As we have previously reported, improved climate-related technical assistance to all levels of government can help limit federal fiscal exposures. Existing federal efforts encourage a decentralized approach to such assistance, with federal agencies incorporating climate-related information into their planning, operations, policies, and programs and establishing their own methods for collecting, storing, and disseminating climate-related data. Reflecting this approach, technical assistance from the federal government to state and local governments also exists in an uncoordinated confederation of networks and institutions. As we reported in our February 2013 high-risk update, the challenge is to develop a cohesive approach at the federal level that also informs action at the state and local levels.[16]

Federal Decision Makers

The Executive Office of the President and federal agencies have many efforts underway to increase the resilience of federal infrastructure and programs. For example, executive orders issued in 2009 and 2013 directed agencies to create climate change adaptation plans which integrate consideration of climate change into their operations and overall mission objectives, including the costs and benefits of improving climate adaptation and resilience with real-property investments and construction of new facilities.[17]

Recognizing these and many other emerging efforts, our prior work shows that federal decision makers still need help understanding how to build resilience into their infrastructure and planning processes. For example, in our May 2014 report, we found that DOD requires selected infrastructure planning efforts for existing and future infrastructure to account for climate change impacts, but its planners did not have key information necessary to make decisions that account for climate and related risks.[18] We recommended that DOD provide further information to installation planners and clarify actions that account for climate change in planning documents. DOD concurred with our recommendations.

Previously, in 2007, we concluded that federal resource management agencies had not made climate change a high priority and did not have specific guidance in place advising their managers on addressing the effects of climate change in their resource management. 19 As a result, we recommended that that the Secretaries of Agriculture, Commerce, and the Interior develop guidance for their resource managers that explains how they expect to address the effects of climate change, and the three departments generally agreed with this recommendation. However, as we found in our May 2013 report, resource managers still struggled to incorporate climate-related information into their day-to-day activities, even with the creation of strategic policy documents and high-level agency guidance.[20]

State and Local Decision Makers

The federal government invests tens of billions of dollars annually in infrastructure projects prioritized and supervised by state and local governments. In total, the United States has about 4 million miles of roads and 30,000 wastewater treatment and collection facilities. According to a 2010 Congressional Budget Office report, total public spending on transportation and water infrastructure exceeds $300 billion annually, with roughly 25 percent of this amount coming from the federal government and the rest coming from state and local governments.[21] However, the federal government plays a limited role in project-level planning for transportation and wastewater infrastructure, and state and local efforts to consider climate change in infrastructure planning have occurred primarily on a limited, ad hoc basis. The federal government has a key interest in helping state and local decision makers increase their resilience to climate change and extreme weather events because uninsured losses may increase the federal government's fiscal exposure through federal disaster assistance programs.

Louisiana State Highway 1 is an example of infrastructure of national importance that is managed by state and local governments. Our April 2013 report on infrastructure adaptation found that according to National Oceanic and Atmospheric Administration estimates, within 15 years, segments of Louisiana State Highway 1 will be inundated by tides an average of 30 times annually due to relative sea level rise. [22] Louisiana Highway 1 is the only road access to Port Fourchon, which services virtually all deep-sea oil operations in the Gulf of Mexico, or about 18 percent of the nation's oil supply. Flooding of this road effectively closes this port. Because of Port Fourchon's significance to the oil industry at the national, state, and local levels, the U.S. Department of Homeland Security, in July 2011, estimated that a closure of 90 days could reduce the national gross domestic product by about $7.8 billion.[23] Figure 4 shows Louisiana State Highway 1 leading to Port Fourchon.

We found in April 2013, that infrastructure decision makers have not systematically incorporated potential climate change impacts in planning for roads, bridges, and wastewater management systems because, among other factors, they face challenges identifying and obtaining available climate change information best suited for their projects.[24] Even when good scientific information is available, it may not be in the actionable, practical form needed for decision makers to use in planning and designing infrastructure. Such decision makers work with traditional engineering processes, which often require very specific and discrete information. Moreover, local decision makers—who, in this case, specialize in infrastructure planning, not climate science—need assistance from experts who can help them translate available climate change information into something that is locally relevant. In our site visits to several locations where decision makers overcame these challenges—including Louisiana State Highway 1—state and local officials emphasized the role that the federal government could play in helping to increase local resilience.

Source: NOAA. | GAO 14-540T.

Figure 4. Louisiana State Highway 1 Leading to Port Fourchon.

Any effective adaptation strategy must recognize that state and local governments are on the front lines in both responding to immediate weather-related disasters and in preparing for the potential longer-term impacts associated with climate change. We reported in October 2009, that insufficient site-specific data—such as local temperature and precipitation projections—complicate state and local decisions to justify the current costs of adaptation efforts for potentially less certain future benefits.[25] We recommended that the appropriate entities within the Executive Office of the President develop a strategic plan for adaptation that, among other things, identifies mechanisms to increase the capacity of federal, state, and local agencies to incorporate information about current and potential climate change impacts into government decision making. USGCRP's April 2012 strategic plan for climate chang e science recognizes this need, by identifying enhanced information management and sharing as a key objective.[26] According to this plan, USGCRP is pursuing the development of a global change information system to leverage existing climate-related tools, services, and portals from federal agencies.

In our April 2013 report, we concluded that the federal government could help state and local efforts to increase their resilience by (1) improving access to and use of available climate-related information, (2) providing officials with improved access to technical assistance, and (3) helping officials consider climate change in their planning processes.[27] As a result, we recommended, among other things, that the Executive Director of USGCRP or other federal entity designated by the Executive Office of the President work with relevant agencies to identify for decision makers the "best available" climate-related information for infrastructure planning and update this information over time, and to clarify sources of local assistance for incorporating climate-related information and analysis into infrastructure planning, and communicate how such assistance will be provided over time.

These entities have not directly responded to our recommendations, but the President's June 2013 Climate Action Plan and November 2013 Executive Order 13653 drew attention to the need for improved technical assistance.[28] For example, the Executive Order directs numerous federal agencies, supported by USGCRP, to work together to develop and provide authoritative, easily accessible, usable, and timely data, information, and decision-support tools on climate preparedness and resilience. In addition, on July 16, 2014, the President announced a series of actions to help state, local, and tribal leaders prepare their communities for the impacts of climate change by developing more resilient infrastructure and rebuilding existing infrastructure stronger and smarter.[29]

We have work under way assessing the strengths and limitations of governmentwide options to meet the climate-related information needs of federal, state, local, and private sector decision makers. We also have work under way exploring, among other things, the risks extreme weather events and climate change pose to public health, agriculture, public transit systems, and federal insurance programs. This work may help identify other steps the federal government could take to limit its fiscal exposure and make our communities more resilient to extreme weather events.

Chairman Murray, Ranking Member Sessions, and Members of the Committee, this concludes my prepared statement. I would be pleased to answer any questions you have at this time.

End Notes

[1] Our past work identified a variety of fiscal exposures—responsibilities, programs, and activities that explicitly or implicitly expose the federal government to future spending. Fiscal exposures vary widely as to source, extent of the government's legal commitment, and magnitude. Further, some of these factors may change over time. For example, the government's response to an event or series of events can strengthen expectations that the government will respond in the same way to similar events in the future. For additional information, see *Fiscal Exposures: Improving Cost Recognition in the Federal Budget*, GAO-14-28 (Washington, D.C.: Oct. 29, 2013).

[2] Melillo, Jerry M., Terese (T.C.) Richmond, and Gary W. Yohe, Eds., 2014: *Climate Change Impacts in the United States: The Third National Climate Assessment*. U.S. Global Change Research Program (Washington D.C.: May 2014). Visit www.globalchange.gov/ for more information about USGCRP. NRC is the principal operating agency of the National Academy of Sciences and the National Academy of Engineering. For more information about NRC, visit www.nationalacademies.org/nrc/index.htm.

[3] U.S. National Academy of Sciences and The Royal Society, *Climate Change: Evidence and Causes (Washington, D.C.: Feb 27, 2014)*.

[4] Congress temporarily increased the borrowing authority for the National Flood Insurance Program by $9.7 billion and provided about $50 billion in appropriated funds for expenses related to the consequences of Superstorm Sandy.

[5] GAO, *High-Risk Series: An Update*, GAO-13-283, February 2013. Every 2 years at the start of a new Congress, GAO calls attention to agencies and program areas that are high-risk due to their vulnerabilities to fraud, waste, abuse, and mismanagement, or are most in need of transformation.

[6] The National Academies, Committee on Increasing National Resilience to Hazards and Disasters; Committee on Science, Engineering, and Public Policy; *Disaster Resilience: A National Imperative* (Washington, D.C., 2012).

[7] GAO, *Climate Change: Future Federal Adaptation Efforts Could Better Support Local Infrastructure Decision Makers*, GAO-13-242 (Washington, D.C.: Apr 12, 2013).

[8] For example, visit www.whitehouse.gov/administration/eop/ceq/initiatives/resilience to access the Climate Change Resilience website maintained by the Council on Environmental Quality (CEQ) within the Executive Office of the President.

[9] See, for example, NRC, Panel on Strategies and Methods for Climate-Related Decision Support, Committee on the Human Dimensions of Global Change, *Informing Decisions in a Changing Climate* (Washington, D.C.: 2009).

[10] GAO, *Climate Change Adaptation: DOD Can Improve Infrastructure Planning and Processes to Better Account for Potential Impacts*, GAO-14-446 (Washington, D.C.: May 30, 2014).

[11] GAO-14-446. Adaptation is defined as adjustments to natural or human systems in response to actual or expected climate change.

[12] GAO-13-242.

[13] GAO, *Climate Change: Various Adaptation Efforts Are Under Way at Key Natural Resource Management Agencies*, GAO-13-253 (Washington, D.C.: May 31, 2013).

[14] Melillo, Jerry M., Terese (T.C.) Richmond, and Gary W. Yohe, Eds., 2014: *Climate Change Impacts in the United States: The Third National Climate Assessment.* U.S. Global Change Research Program (Washington D.C.: May 2014).

[15] Visit http://www.gao.gov/key_issues/wildland_fire_management/issue_summary to access a summary of wildland fire management issues and related reports on GAO's Key Issues website. See also Congressional Research Service, *Wildfire Management: Federal Funding and Related Statistics,* R43077 (March 5, 2014).

[16] GAO-13-283.

[17] Executive Order 13514, *Federal Leadership in Environmental, Energy, and Economic Performance* (Oct. 5, 2009). Executive Order 13653, *Preparing the United States for the Impacts of Climate Change* (Nov. 1, 2013).

[18] GAO-14-446.

[19] GAO, *Climate Change: Agencies Should Develop Guidance for Addressing the Effects on Federal Land and Water Resources,* GAO-07-863 (Washington, D.C.: Aug. 7, 2007).

[20] GAO-13-253.

[21] Congressional Budget Office, *Public Spending on Transportation and Water Infrastructure*, Pub. No. 4088 (Washington, D.C.: November 2010).

[22] GAO-13-242.

[23] Department of Homeland Security, National Infrastructure Simulation and Analysis Center, Risk Development and Modeling Branch, Homeland Infrastructure Threat and Risk Analysis Center, Office of Infrastructure Protection, in collaboration with the National Incident Management Systems and Advanced Technologies Institute at the University of Louisiana at Lafayette, *Louisiana Highway 1/Port Fourchon Study* (July 15, 2011).

[24] GAO-13-242.

[25] GAO, *Climate Change Adaptation: Strategic Federal Planning Could Help Government Officials Make More Informed Decisions,* GAO-10-113 (Washington, D.C.: Oct. 7, 2009).

[26] USGCRP, National Global Change Research Plan 2012-2021: *A Strategic Plan for the U.S. Global Change Research Program* (Washington D.C.: April 2012).

[27] GAO-13-242.

[28] More information on the June 2013 Climate Action Plan and Executive Order 13653 can be found at www.whitehouse.gov/administration/eop/ceq/initiatives/resilience.

[29] Visit www.whitehouse.gov/the-press-office/2014/07/16/fact-sheet-taking-action-support-state-local-and-tribal-leaders-they-pre for more information on the resilience efforts announced on July 16, 2014.

In: Economic Costs of Inaction on Climate Change ISBN: 978-1-61728-031-3
Editor: Cheryl Griffin © 2014 Nova Science Publishers, Inc.

Chapter 4

TESTIMONY OF SHERRI W. GOODMAN, EXECUTIVE DIRECTOR, CNA MILITARY ADVISORY BOARD. HEARING ON "THE COSTS OF INACTION: THE ECONOMIC AND BUDGETARY CONSEQUENCES OF CLIMATE CHANGE"[*]

Chairman Murray, Ranking Member Sessions, and Committee members, I thank you for inviting me to testify today.

INTRODUCTION: CNA MILITARY ADVISORY BOARD: MILITARY LEADERSHIP AND CLIMATE RISKS

I am Sherri Goodman, and I am privileged to serve as the founder and Executive Director of CNA's Military Advisory Board—MAB for short. In this capacity, I am here today representing not only my views on the national security implications of climate change, but also the collective wisdom of the 16 Admirals and Generals who serve on CNA's MAB.

This board first convened in 2006 to look at pressing national security issues, including climate change. Our first report, published in 2007, identified climate change as a threat multiplier, especially in fragile regions of the globe. Since that

---
[*] This is an edited, reformatted and augmented version of testimony presented July 29, 2014 before the Senate Budget Committee.

first report, we have had over 30 Generals and Admirals serve on the on the board, collectively with more than one thousand years of experience in evaluating security threats and mitigating risks. Our most recent report, which I would like to submit for the Record, identifies the accelerating risk of climate change and observes that in some circumstances climate change has, and increasingly will, serve as a catalyst for conflict.

To explain how the accelerating risk of climate change may impact the U.S. budget from a national security perspective, I will focus on the following four areas:

- First, global trends that will contribute to instability around the world;
- Second, the Arctic;
- Third, military readiness; and,
- Fourth and finally, U.S. National Power.

My discussion today is informed by the MAB and reflects its members' most recent findings, but what follows are my own views and observations.

I. GLOBAL TRENDS: ACCELERATING RISKS

In the seven years that have passed since our initial assessment, we have witnessed more frequent and/or intense weather events, including heat waves, sustained heavy downpours, floods in some regions, and droughts in other areas. Nine of the ten costliest storms to hit the United States have occurred in the past 10 years, including Hurricane Katrina and Superstorm Sandy. Speaking for the MAB, we assess that the nature and pace of observed climate changes—and an emerging scientific consensus on their projected consequences—pose severe risks for our national security.

Having served for eight years as Deputy Undersecretary of Defense for Environmental Security, and for eight more years as Executive Director of the MAB, I have learned how our senior military leaders approach risk and uncertainty. To our military leaders, managing risk is seldom about dealing with absolute certainties but, rather, involves careful analysis of the probability of an event and the consequences, should the event occur. When it comes to our national security, very low probability events with potentially dire consequences often deserve consideration and contingency planning. Military leaders evaluate the probability and possible consequences of events in determining overall risk. Today, the risks posed by predicted climate change represent even graver potential than they did seven years ago, and require action today to reduce increased risks and potential impacts tomorrow.

A. Four Important Global Trends

Four notable global trends will exacerbate these accelerating risks. First is *global population growth:* One-half billion people have been added to the world's population since the MAB completed its first report in 2007 and another one-half billion will be added by 2025. Most of this growth is in Africa and Asia, two of the areas projected to be hardest hit by climate change.

The second trend is *urbanization:* Nearly half of the world now lives in urban areas, with 16 out of 20 of the largest urban areas situated near coastlines. The result is that more of the world's population is at risk from extreme weather events, sea level rise, and storm surge.

The third trend is a global increase in the middle class, with an accompanying *growth in demand for food, water, and energy.* The National Intelligence Council predicts that by 2030, demand for food will increase by 35 percent, fresh water by 40 percent, and energy by 50 percent.[1] Another 2012 assessment by the U.S. intelligence community found that water challenges will likely increase the risk of instability and state failure, exacerbate regional tensions, and divert attention from working with the United States and other key allies on important policy objectives.[2]

The fourth and final trend is that the world is becoming more politically complex and *economically and financially interdependent, which means that security risks* to any one region of the world cannot be examined in isolation.

B. Accelerating Risks Around the World Affect U.S. National Security

The world around us is changing. In recent years, scientists have observed changing weather patterns manifest by prolonged drought in some areas and heavier precipitation in others. In the last few years, we have seen unprecedented wildfires threaten homes, habitats, and food supplies—not only across the United States, but also across Australia, Europe, Central Russia, and China. Low-lying island nations are preparing for complete evacuation to escape rising sea levels. Globally, recent prolonged drought has acted as a factor driving both spikes in food prices and mass displacement of populations, each contributing to instability and eventual conflict. For example, the MAB notes that drought conditions in Russia and China, and subsequent global wheat shortages, contributed to higher food prices in Northern Africa and may have helped catalyze and sustain the Tunisian and Egyptian uprisings in 2011. Similarly, in Syria, five years of drought decimated farms and forced millions to migrate to urban areas. In overpopulated cities, these refugees found little in the way of jobs and were quickly disenfranchised with the government. The ongoing strife in Syria has been exacerbated by drought and rural to urban migration. In this way, climate change has worsened stress in a region already torn by political and ethnic tensions, serving as a catalyst for conflict.

We are concerned about the projected impacts of climate change over the coming decades on those areas already stressed by water and food shortage and poor governance. Such areas span the globe, and they present the greatest short-term threats. In the longer term, the areas that are at the greatest risk are those exposed to rising sea levels. There will be only so much we can do to keep the sea out, and in some areas the sea will not flow over the walls we build, it will flow under or around, making the land and aquifers unusable. Low-lying islands in the Pacific and great deltas, including the Mekong, the Ganges in Bangladesh, the Nile Delta in Egypt, the Mississippi Delta, and whole regions like the Everglades are increasingly at risk of being unable to support the populations that live there. Sea water inundation will drastically cut food production in many of these areas and cause millions to lose their ability to live on these retreating arable lands. In these areas and in others, migration could become a larger method of adaptation.

II. Accelerating Climate Risks to the US Homeland

A. The Arctic Region Is Rapidly Changing—and the U.S. Needs to Prepare

The Arctic is a region experiencing rapid change. Over the past few years, we have seen an almost exponential rise in activity in the Arctic: more shipping, more resource extraction, and more posturing for control over the region's vast resources. The international community is not yet prepared to respond to an accident or disaster that could occur with increasing shipping and energy exploration in this fragile region, with its limited infrastructure and extreme operating conditions. In the Arctic, climate change challenges could serve as a catalyst for increased international cooperation.

While serving as Deputy Undersecretary of Defense in the aftermath of the Cold War, the U.S. worked with Norway, Russia, and others to manage waste streams from decommissioned Russian nuclear submarines, including some that had been dumped into the Kara Sea, north of the Arctic Circle. In helping Russia safely manage waste streams from nuclear ship operations, I became acutely aware of the unique Arctic environment. With increased shipping and greater opportunities for extraction of resources, the risk for a man-made crisis or disaster, such as a major oil spill, is rising.

A report of April 2014 on *Responding to Oil Spills in the U.S. Arctic Marine Environment* by the National Research Council finds that a spill in the Arctic, similar in size to that of the Deepwater Horizon accident in 2010 would have a devastating impact and the effects would last for decades. During the Deepwater Horizon response, technicians had continuous access to the sight and it still took three months to stop the leak and two more to seal it. Depending on the time of year, that level of access would not be possible in the Arctic. For the Horizon accident, responders mobilized hundreds of privately owned boats to control, contain, and ultimately disperse the surface oil slick. This simply would not be possible in the Arctic. As stated by the National Research Council, "In the presence of lower water temperature or sea ice, the processes that control oil weathering – such as spreading, evaporation, photo oxidation, emulsification, and natural dispersion—are slowed down or eliminated and capture of oil in new ice can make the spill lasts for months."[3] Simply put, the world is not prepared to respond to a major accident in the Arctic, be it from drilling or the transshipment of oil.

Some great work has been done across the U.S. government in putting together plans for increased future operations in the Arctic. The problem, though, is that the increase is happening *now*. Seventy-three ships sailed through the Northwest Passage in 2013, up from just four in 2007. Preparations for energy exploration are well underway—the Russians have already staked out claims over potential reserves by planting a flag on the deep seabed near the North Pole. My colleagues on the MAB warn that today, the U.S. does not have the communications equipment, navigation aids, or sufficient hardened-hull ships to respond to natural or man-made disasters in that fragile area or to protect our vital interests in the region. In other words, we are not prepared in the short term for the rate of increase in Arctic activity. We must invest today in increasing our capability and capacity.

B. Growing Awareness of Climate Risks and Planning in the U.S.

On the positive side, we have seen increased awareness of climate risks in communities around the U.S., and constructive planning underway in various regions. Two examples are worth noting.

Pacific Northwest Provides a National Model for Action

On June 4, 2014, I participated in a symposium in Seattle convened by the Henry M. Jackson Foundation and the U.S. Department of Energy's Pacific Northwest National Laboratory on "National Security and Climate Change." Bringing together local and national leaders and practitioners, this effort used the CNA Military Advisory Board's recent report as a launching point to explore how to address climate security risks in the Pacific Northwest. The symposium report states:

> Many communities in the Pacific Northwest are serious about addressing climate change and national security threats. Climate scientists have reported that the Northwest region of the country will likely experience increasing wildfires from decreased snowpack, increasing storms leading to flooding, and rapid ocean acidification.
>
> These changes pose numerous economic and safety challenges, such as alterations of salmon spawning patterns and increased risks of rockslides that threaten both infrastructure and human life. Furthermore, sea level rise also places critical infrastructure such as railroads and ports at risk due to their low elevation. Federal, state, and regional governments will be forced to deploy resources and manpower to respond.
>
> Local governments are using these findings in future planning. For example, the bipartisan King County Council, which covers the greater Seattle region, called for the development of a strategic climate action plan. The groundbreaking plan provides a blueprint for carbon mitigation and adaptation that could be used as a national model for other localities.
>
> Utilities in the region are also planning for impacts, including the ability to serve military bases.

Hampton Roads, Virginia: Partnerships to Manage Sea Level Rise

A second example is Hampton Roads, Virginia, where sea level rise is being jointly addressed by the military and the local community. (The CNA MAB report features Hampton Roads as a case study on page 25.) Rising sea levels, natural subsidence, and storms pose risks to the many military facilities, related commercial shipyards and suppliers, and the community in this critically important region. The area has hundreds of miles of waterfront from three major rivers that all flow into the Chesapeake Bay. DOD realizes that sea level rise will affect both the Hampton Roads installations and the surrounding civilian community. Working with other federal, state, and local agencies, and Old Dominion University, DOD has launched an aggressive effort to develop plans and measures to sustain the vital missions of this region and protect the large surrounding community.

III. INCREASING IMPACTS ON MILITARY READINESS

The MAB finds that projected changes in climate will have three major impacts on the military: more demand, challenges to readiness, and new and harsher operating environments.

The MAB expects to see an increased demand for forces across the full spectrum of operations. Domestically, responses to extreme weather events and wildfires in the U.S. will increase demand for National Guard and reserves. The frequency, severity, and probability that these events could happen simultaneously will also likely increase demand for active duty forces to provide defense support of civil authorities (DSCA). This concerns us because, in a leaner military, many of our capabilities reside in the Guard and reserve, and if they are being used domestically, they are less available to respond to worldwide crises.

Globally, there will be increased demands for humanitarian response and disaster relief in response to extreme weather events. Witness more than 13,000 military troops who responded to Typhoon Haiyan in the Philippines late last year.

In addition to more demand, which in itself will stress readiness, our bases will be increasingly at risk. Our bases are vulnerable to sea level rise and extreme weather, including drought and, in the future, increased precipitation in the form of rain and snow. Drought and the threat of wildfires have already caused live fire training restrictions on major training ranges in Texas and Southern California and, earlier this year, a wildfire at Camp Pendleton shut down training and caused partial evacuation of the base. It is not just the bases that are vulnerable, but also the surrounding communities that house and support the military. If our sailors, soldiers, airmen, and marines can't get to their bases, because the roads are flooded, then we can't maintain the readiness of the force.

Finally, the impacts of climate change will cause the military to be deployed to harsher environments. Higher temperatures have and will continue to stress equipment and people.

IV. NATIONAL POWER AFFECTED BY CLIMATE RISKS

The final area I want to cover is how climate change will impact the elements of national power.

National security is more than just having a strong or capable military. America's security is determined by multiple elements of National Power: diplomacy, infrastructure, military, and economic assets, to name just a few. When deployed strategically, they can constitute "smart power." On the vulnerability side, National Power can also be assessed by degradations to these assets or systems.

A. Strain on Military Readiness and Base Resilience

As mentioned earlier, the projected impacts of climate change could be detrimental to military readiness; strain base resilience, both at home and abroad; and limit our ability to respond to future demands. More forces will be called on to respond in the wake of extreme weather events at home and abroad, limiting their ability to respond to other contingencies. Projected climate change will make training more difficult, while at the same time, will put at greater risk critical military logistics, transportation systems, and infrastructure—both on and off base.

As coastal regions become increasingly populated and developed, more frequent or severe storms increasingly will threaten vulnerable populations in these areas and increase the requirements for emergency responders. Simultaneous or widespread extreme weather events and/or wildfires, accompanied by mass evacuations, and degraded critical infrastructure could outstrip local and federal government resources, and require the increased use of military and private sector support.

B. Risks to Critical Infrastructure

The nation depends on critical infrastructure for economic prosperity, safety, and the essentials of everyday life. All 16 critical infrastructure sectors identified in the Department of Homeland Security planning directives will be impacted by our changing climate. We are already seeing how extreme heat is damaging the national transportation infrastructure such as roads, rail lines, and airport runways. Moreover, much of the nation's energy infrastructure—including oil and gas refineries, storage tanks, power plants, and electricity transmission lines—is located in coastal floodplains, where it is increasingly threatened by more intense storms, extreme flooding, and rising sea levels. Projected increased temperatures and drought across much of the nation will strain energy systems (with more demand for cooling) and increase water stress. Since much of the critical infrastructure is owned or operated by the private sector, government and the private sector will need to work closely together to develop solutions to address the full range of these challenges.

C. Economic Risks

The MAB holds that a strong economy is critical to national security. In June of 2014, Hank Paulson, who served as Secretary of the Treasury under President George W. Bush, co-chaired a panel of business leaders and economic experts and published the report, "Risky Business: the Economic Risks of Climate Change in the United States."

The report stated:

Our key findings underscore the reality that if we stay on our current emissions path, our climate risks will multiply and accumulate as the decades tick by. These risks include:
Large-scale losses of coastal property and infrastructure:

- Property losses from sea level rise are concentrated in specific regions of the U.S., especially on the Southeast and Atlantic coasts, where the rise is higher and the losses far greater than the national average.

Extreme heat across the nation—especially in the Southwest, Southeast, and Upper Midwest—threatening labor productivity, human health, and energy systems:

- By the middle of this century, the average American will likely see 27 to 50 days over 95°F each year—two to more than three times the average annual number of 95°F days we've seen over the past 30 years. By the end of this century, this number will likely reach 45 to 96 days over 95°F each year on average....
- Demand for electricity for air conditioning will surge in those parts of the country facing the most extreme temperature increases, straining regional generation and transmission capacity and driving up costs for consumers.

Shifting agricultural patterns and crop yields, with likely produce gains for Northern farmers offset by losses in the Midwest and South:

- As extreme heat spreads across the middle of the country by the end of the century, some states in the Southeast, lower Great Plains, and Midwest risk up to a 50% to 70% loss in average annual crop yields (corn, soy, cotton, and wheat), absent agricultural adaptation....
- Food systems are resilient at a national and global level, and agricultural producers have proven themselves extremely able to adapt to changing climate conditions. These shifts, however, still carry risks for the individual farming communities most vulnerable to projected climatic changes."[4]

Conclusion: The Time Is Now

In sum, projected climate change may cause increased instability around the world; we are not prepared for the pace of climate change as evidenced by our limited capability and capacity to respond to the opening of the Arctic; climate change will likely impact our military readiness and support systems as well as lead to increased demand for forces, both at home and abroad, and finally climate change will impact elements of our national power here at home. Let me leave you with these comments by the 16 Generals and Admirals who authored our most recent report:

> At the end of the day, we validate the findings of our first report and find that in many cases the risks we identified are advancing noticeably faster than we anticipated. We also find the world becoming more complex in terms of the problems that plague its

various regions. Yet thinking about climate change as just a regional problem or—worse yet—someone else's problem may limit the ability to fully understand its consequences and cascading effects. We see more clearly now that while projected climate change should serve as catalyst for change and cooperation, it can also be a catalyst for conflict.

We are dismayed that discussions of climate change have become so polarizing and have receded from the arena of informed public discourse and debate. Political posturing and budgetary woes cannot be allowed to inhibit discussion and debate over what so many believe to be a salient national security concern for our nation.[5]

In their foreword to the CNA MAB report, former Secretary of Homeland Security Michael Chertoff and former Secretary of Defense Leon Panetta summarized our most important message for the Committee: "The update serves as a bipartisan call to action. It makes a compelling case that climate change is no longer a future threat—it is taking place now. . . . [A]ctions to build resilience against the projected impacts of climate change are required today. We no longer have the option to wait and see."[6]

End Notes

[1] National Intelligence Council, *Global Trends 2030: Alternative Worlds* (Washington, DC: Office of the Director of National Intelligence, December 2012).

[2] Office of the Director of National Intelligence, *Global Water Security*. Intelligence Community Assessment ICA 2012-08 (Washington, DC: Office of the Director of National Intelligence, February 2, 2012).

[3] National Research Council. Responding to Oil Spills in the U.S. Arctic Marine Environment. Washington, DC: The National Academies Press, 2014.

[4] "Risky Business; the Economic Risks of Climate Change in the United States." Retrieved on 25 July, 2014 from http://riskybusiness.org/report/overview/executive-summary

[5] CNA Military Advisory Board, *National Security and the Accelerating Risks of Climate Change* (Alexandria, VA: CNA Corporation, May 2014), p. iii.

[6] Ibid., p. i.

In: Economic Costs of Inaction on Climate Change ISBN: 978-1-61728-031-3
Editor: Cheryl Griffin   © 2014 Nova Science Publishers, Inc.

Chapter 5

TESTIMONY OF BJØRN LOMBORG, DIRECTOR, COPENHAGEN CONSENSUS CENTER. HEARING ON "THE COSTS OF INACTION: THE ECONOMIC AND BUDGETARY CONSEQUENCES OF CLIMATE CHANGE"[*]

SUMMARY

Global warming is real, but a problem, not the end of the world Climate is not the only cost – climate policy also adds costs

Even the smartest climate policy will likely have almost as high total costs as inaction this century

A more politically plausible climate policy will have much higher cost than inaction this century

To answer the committee's question on the US cost of inaction and action for climate change:

- The total, discounted cost of inaction on climate change over the next five centuries is about 1.2% of total discounted GDP.

---

[*] This is an edited, reformatted and augmented version of testimony presented July 29, 2014 before the Senate Budget Committee.

- The cumulative cost of inaction towards the end of the century is about 1.8% of GDP
- While this is not trivial, it by no means supports the often apocalyptic conversation on climate change.
- The cost of inaction by the end of the century is equivalent to losing one year's growth, or a moderate, one-year recession.
- The cost of inaction by the end of the century is equivalent to an annual loss of GDP growth on the order of 0.02%.
- However, policy action as opposed to inaction, also has costs, and will still incur a significant part of the climate damage. Thus, with extremely unrealistically optimistic assumptions, it is possible that the total cost of climate action will be reduced *slightly* to 1.5% of GDP by the end of the century.
- It is more likely that the cost of climate action will end up costing upwards of twice as much as climate inaction in this century - a reasonable estimate could be 2.8% of GDP towards the end of the century.
- Thus, for the first half century, it is absolutely certain that any climate action will have greater total costs than inaction. For the second half of the century it is very likely that any realistic climate action will have greater costs than climate inaction.
- While it is possible to design clever, well-coordinated, moderate climate policies that will do more good than they will cost, it is much more plausible that total costs of climate action will be more expensive than climate inaction.
- To tackle global warming, it is much more important to dramatically increase funding for R&D of green energy to make future green energy much cheaper. This will make *everyone* switch when green is cheap enough, instead of focusing on inefficient subsidies and second-best policies that easily end up costing much more.
- It is likely that the percentage cost to the US budget is in the same order of magnitude as that of the percentage costs to the US economy.

WHAT IS THE COST OF INACTION AND ACTION?

Is global warming happening? Yes. Man-made global warming is a reality and will in the long run have overall, negative impact.

It is important to realize that many economic models show that the overall impact of a moderate warming (1-2oC) will be beneficial whereas higher temperatures expected towards the end of the century will have a negative net impact.[1] Thus, as indicated in Figure 1, global warming is a *net benefit* now and will likely stay so till about 2070, after which it will turn into a net cost.

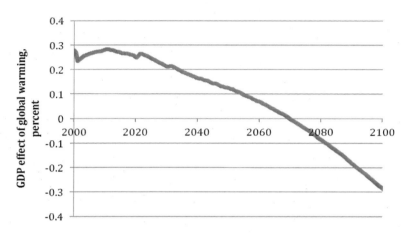

Figure 1. Net benefit or cost of global warming. Benefit is positive, cost is negative.[2]

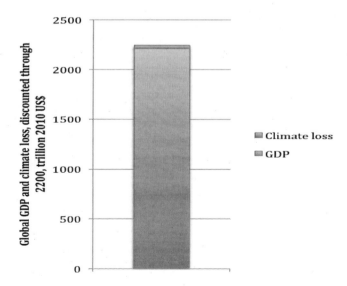

Figure 2. Global, total, discounted GDP through 2200, and climate loss.[3]

How important is global warming? To get a sense of the importance of global warming, take a look at the total impact of damage compared to the cumulated consumption using the discount rates from Nordhaus' 2010 DICE model. The total, discounted GDP through the year 2200 (almost the next two centuries) is about $2,212 trillion dollars. The total damage is estimated at about $33 trillion or about 1.5% of the total, global GDP, as indicated in Figure 2.

This means that while the global warming impact is *not* zero but *negative*, it does *not* signify the end of the world, either. It is a problem that needs to be solved.

What is the impact of global warming on the US economy? There are a number of integrated climate models. I'll here use Nordhaus' RICE model[4] The model contains 12 regions, including the US, China and the EU, an economic sector and geophysical sectors, linking the economy and climate impacts like sea level rise. It has a equilibrium climate sensitivity of 3.2_oC, a bit above average, expecting 3.4_oC temperature rise by 2100 in the base scenario.

Remember also, that the costs of the risks of abrupt and catastrophic climate change are included in the damage estimates in the RICE model.

The RICE model shows instant damages from temperature, making it more pessimistic than most estimates, as referenced above. Moreover, the model shows a 1.95% GDP loss in 2075 from unrestricted global warming at 1.95_oC. The IPCC found that the cost of 2_oC higher temperatures would be 0.2-2% of income.[5] This means that the RICE model, if anything, is at the high end cost estimates of the integrated models.

The RICE model show the total, discounted GDP for the US across the next 5 centuries is about $842 trillion (2005$), but this will be reduced by about $10 trillion from cumulative impacts from global warming, as indicated in Figure 3. This means that the total damages from unmitigated global warming for the US is on the order of 1.2%.

This indicates, as has often been pointed out, that the US is *less* vulnerable to climate change, compared to many other regions (especially the poorer countries).

Moreover, it emphasizes that while the global warming impact is *a net negative* for the US, it is in no way a catastrophe, either.

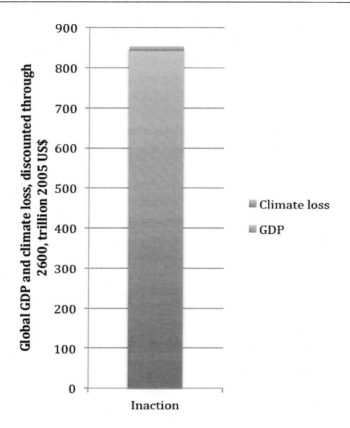

Figure 3 Total US, discounted GDP through 2500, and climate loss.[6]

How much will global warming directly impact the federal budget? I know of no direct estimate of the total impact of global warming on the federal budget. Consequently, I will here assume that the main impact of global warming on the federal budget will be a reduction in total revenue, in line with the reduction in US GDP due to global warming (expecting unchanged taxes). On the one hand, because not all damages included in the RICE model will be translated into actual GDP losses, this may be an over-estimate. On the other hand, it is likely that parts of the costs of global warming will be borne disproportionately by the federal government. Thus, in total, it is likely that the loss estimate from GDP from the RICE model translates directly into the negative impact on the federal budget. In the following discussions I'll treat the impact to the US economy and the federal budget as similar percentages (although of course, of a different base, since the US GDP is about $16 trillion, and the federal budget is about $3 trillion).

That means that the total direct impact on the US federal budget is likely to be a reduction of about 1.2% across the next centuries.

However, this is not actually the avoidable impact from climate, since some climate impact will happen no matter what we do. The internationally most ambitious target (which is probably almost out of reach) is the 2oC goal. Figure 4 shows the cost of unmitigated global warming in the upper line, reaching a US cost of 1.8% of GDP by 2100. The lower, 2oC line shows a cost that is almost indistinguishable for the first decades, leveling off just below 0.6% of GDP by 2100. Thus, the avoidable global warming is the area between the two lines, or about 1.2% GDP by 2100.

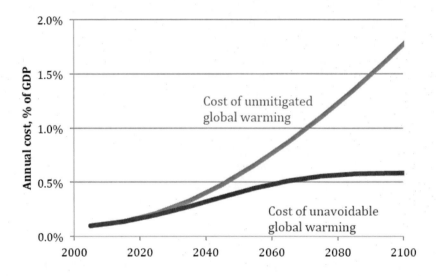

Figure 4. US cost for each year, in % of US GDP that year. Upper line shows the cost of unmitigated global warming. Lower line shows the unavoidable cost of global warming, if all nations achieve the most efficient policies to reach the 2oC target. All calculations from RICE.

The RICE model show the total, discounted GDP impact of global warming for the US across the next 5 centuries is $10 trillion, as mentioned above, while the cost of the unavoidable global warming is about $3 trillion. This means that the total avoidable damages from global warming for the US is on the order of 0.8%.

With similar reasoning as above, it seems likely that the total avoidable impact on the US federal budget will be in the order of 0.8% of GDP.

How much will global warming indirectly impact the US economy? It is important to remember that the cost of global warming is not the only impact on the US economy or the federal budget. Any climate policy enacted to (partially) counter global warming will also carry both costs and benefits. These will indirectly, through policy, impact both the US economy and the federal budget.

The 2oC policy. Consider the world implementing the widely promised (but fairly unlikely) 2oC implemented in the most efficient way possible. This would entail a single, global, uniformly imposed carbon tax, which would increase rapidly through the century. In the RICE model, the indication is that the global carbon tax would have had to be $19/ton CO_2 in 2010, and would have to be $26 in 2015 and $16 in 2020, about $170 in 2055 and $296 in 2105.[7]

To give an indication, this would add ¢22 to a gallon of gasoline about now and $3.40 to a gallon of gasoline in 2085, across the world, including the poorest places on earth.

This is already politically very unlikely to happen. Moreover, the cost is likely a low estimate. Another survey of a 8 global energy models showed the 2oC target might cost in the order of 12.9% of GDP by the end of the century, leading to carbon taxes of four times the RICE model at $4004 per ton CO_2.[8]

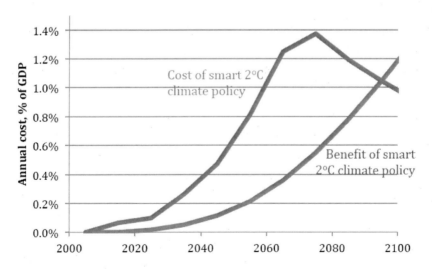

Figure 5. US cost and benefit for each year, in % of US GDP that year of 2oC efficient climate policy. Blue line shows net benefit (avoided costs) from less global warming. Red line shows extra cost. All calculations from RICE.

The important point to realize here is that the costs to the US fall heavily in the early part of the period whereas the benefits tend to come later. This is a standard finding for all climate models and all climate policies.

Here, the cost to the US economy will run upwards of 1.4% of GDP in the second half of the century or about $600 billion in annual costs vs. $250 billion in avoided damages.

Despite everyone else including China and India also implementing similarly expensive climate policies, the US costs will outweigh the benefits for the US from this global policy until the early 2090s, although the benefits will clearly outweigh the costs in the 22nd century and beyond.

With Nordhaus' discounting this climate policy is actually still seen as socially beneficial, because the benefits from future centuries sufficiently outweigh the net cost in this century. The avoided damages run to almost $7 trillion, whereas the policy costs a bit more than $4 trillion. The numbers are almost similar with a traditional 3% discount rate, but with a 5% discount rate, the total policy costs are more than twice the benefits.

Moreover, it seems unlikely that other countries would enact this sort of policy. The annual costs for China would in 2065 be $863 billion annually, with benefits of just $170 billion.

The 'optimal' climate policy. The optimal policy in the RICE model is estimated as the climate policies coordinated and enacted by all nations starting in 2010 that maximize global economic welfare across the next six centuries.

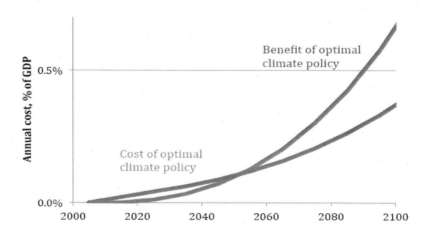

Figure 6. US cost for each year, in % of US GDP that year of optimal climate policy. Blue line shows net benefit (avoided costs) from less global warming. Red line shows extra cost. All calculations from RICE.

The costs and benefits for the US can be seen in Figure 6. Again, the costs outweigh the benefits for the first half-century, but the benefits significantly outweigh the costs for the coming centuries.

This policy is less politically prohibitive, since it requires a lower carbon tax. In the RICE model, the indication is that the global carbon tax would have had to be $9/ton CO_2 in 2010, $12 in 2015 and $16 in 2020, about $50 in 2050 and $130 in 2100.[9] In terms of gasoline, this would have added about 8 on a gallon in 2010 globally, ¢18 in 2020, about ¢40 in 2050 and $1.14 in 2100.

This policy is a net benefit, and quite substantial. With Nordhaus' discounting, it costs the world $1.5 trillion, but avoids climate damages worth $5 trillion. With 5% discount rate, it is still a slight net benefit.

Yet, actually seeing this policy enacted is wholly unrealistic, as Nordhaus acknowledges.[10] It requires policies that would be coordinated across the entire world, with carbon taxes imposed even on the poorest nations. For instance, the costs for China would remain higher than the Chinese benefits until after 2080, making this a very hard political sell.

As Nordhaus points out, the costs up till mid-century are five times higher than the benefits:

> Abatement costs are more than five times the averted damages. For the period after 2055... however, the ratio is reversed: Damages averted are more than four times abatement costs. Asking present generations— which are, in most projections, less well off than future generations—to shoulder large abatement costs would be asking for a level of political maturity that is rarely observed.

Importantly, the optimal policy will avoid very little of global warming impacts in the 21st century. Figure 7 shows the total damages for both action and inaction.

The damages for inaction (business-as-usual) is just the climate damage from Figure 4, with a cost of about 0.14% of GDP now, and a cost of 1.8% of GDP in 2100. The cost of the optimal, globally coordinated climate policy is the cost of climate policies and the residual negative climate impact. It starts out slightly higher at a cost of 0.16% of GDP now and with a cost of 1.4% of GDP in 2100.

Remembering this is a wholly unrealistic policy to be implemented and be implemented well, the most optimistic statement that can be made on the cost of action and inaction on climate change for the US in the 21st century is that there is little difference. Starting out more expensively, even the optimal climate policy will incur nearly as much cost as no action at all, at 1.4%

instead of 1.8% of GDP by the end of the century. As will be apparent below, this is an extremely and unrealistically rosy assessment.

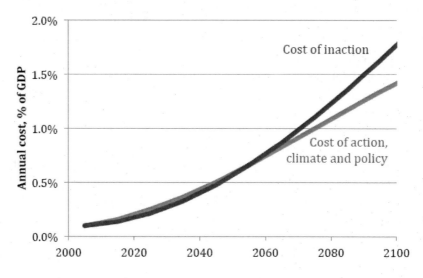

Figure 7. Total cost of climate impact and climate policy for the US. Dark blue line shows the total cost of inaction. Light blue line shows the total cost of smartest, globally coordinated action, both from policy and residual climate damage. All calculations from RICE.

Mostly rich world, ambitious reductions. Both India and China have defended their right to keep their emissions increasing. It is unlikely that they or the rest of the developing, mostly very poor countries will substantially reduce their emissions anytime soon. Nordhaus develops a scenario with rich countries (US, EU, Japan, Russia and the the rest of the rich countries) engage in strong emissions reductions but where the developing countries only participate in the 22^{n}d century.[11] On the current set of policies from both rich and poor countries, this scenario seems a lot more realistic.

In this scenario, the costs are greater than the optimal policy for the rich countries, because they have offered to cut much, much more. This is evident in the EUs professed approach to cut emissions at least 80% below 1990 levels by 2050, and in similar statements from the current US administration.

The benefits, however, are smaller, because many of the biggest emitters are not included. This is readily evident in Figure 8, where China now emits almost twice what the second-largest emitter, the US, does. Of course, China, India and the other poor country emitters will still experience a net benefit in lower climate damages due to the generous reductions from the rich countries.

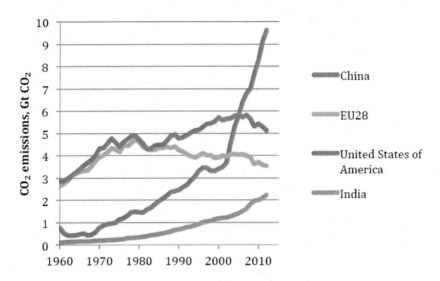

Figure 8. CO_2 emissions from the leading four emitters, China, US, EU and India, 1960`2012.[12]

Nordhaus estimate the future US reductions from the 2009 US climate bill that was passed by the House but not the Senate. In this scenario, the US will by mid-century have reduced its emissions some 75% below what they would otherwise have been.

The climate policy costs for the US will not be trivial. Assuming a full trading zone between all participants, the annual policy costs will run to $145 billion by mid-century and some $250 billion by the end of the century, or about 0.4% of GDP. The full trading assumption is rather unrealistic, as trading has generally been only weakly implemented and often only for small parts of the emissions spectrum. The more realistic cost with a no-trade assumption shows the US costs at about twice the annual cost at $280 billion by mid-century and $400 billion by the end of the century.

We can check the reasonableness of these costs by looking at the well-modeled costs of the EU climate policy to 2020.[13] The average cost by 2020 from 6 models runs to €209 billion or about $280 billion per year (1.3% of GDP). The Nordhaus model (admittedly doing a much more simplified analysis) finds the cost at less than $5 billion, even without trade, suggesting that the RICE estimates are certainly not exaggerated.

However, a consistent result from the studies of the EU climate policy is that real climate policies are often poor, second-best policies, with a mish-mash of regulation of different sectors and regions. The most pertinent summary of the Stanford Energy Modeling Forum's assessment of the EU policies finds:

> Second-best policies increase costs. A policy with two carbon prices (one for the ETS, one for the non-ETS) could increase costs by up to 50%. A policy with 28 carbon prices (one for the ETS, one each for each Member State) could increase costs by another 40%. The renewables standard could raise the costs of emissions reduction by 90%. Overall, the inef!ciencies in policy lead to a cost that is 100–125% too high.[14]

Thus, it is very likely that a more realistic estimate of costs will be a bit above twice the optimal estimate. For the RICE model, that means that the US costs of an ambitious climate policy will more likely incur annual costs of about half a trillion by mid-century and some $800 billion by the end of the century.

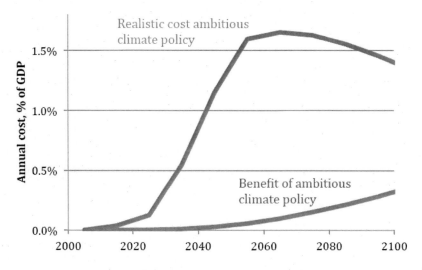

Figure 9. US cost and benefits for each year, in % of US GDP that year of realistic, ambitious climate policy ("Copenhagen Accord with only rich countries," no trade and 2x policy costs). Blue line shows net benefit (avoided costs) from less global warming. Red line shows policy costs. All calculations from RICE.

The overview of the 21st century is available in Figure 9. The policy cost is vastly greater than the avoided climate damages, with costs running above 1.5% of GDP (about similar to what the moderate EU climate efforts will cost the EU by 2020), while benefits run between 0.1% and 0.3% in the second half of the century.

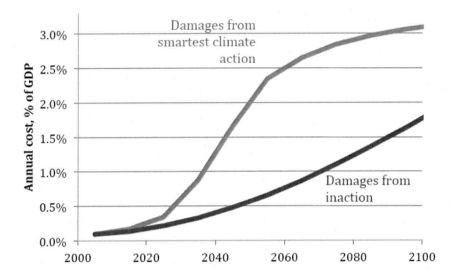

Figure 10. Total damages from climate impact and climate policy costs for the US, in % of US GDP that year. Dark blue line shows the total cost of inaction. Light blue line shows the total cost of realistic, ambitious climate action. All calculations from RICE.

Again, it is important to emphasize that such an ambitious climate policy does not reduce total impacts to the US economy or the federal budget, but actually dramatically increase the total cost, as is evident in Figure 10. In such a situation the US would have to both suffer significant costs from only slightly reduced climate change while incurring even higher policy costs.

Figure 11 answers the committee's question on the costs of climate inaction and climate action. The costs of inaction rise through the century to about 1.8% of GDP in 2100. With extremely unrealistically optimistic assumptions, it is possible that the total cost of climate policy action will be reduced *slightly* to 1.5% of GDP by the end of the century. With more likely assumptions, the cost of climate action will end up costing upwards of twice as much as climate inaction in this century, or about 3.1% of GDP towards the end of the century. No matter what, the cost of action is higher than the cost of inaction in the first half of the century.

Another way to see look at the cost of action and inaction is to look at the total, discounted cost of global warming and global warming policy on the 21st century in Figure 12. The cost for the unrealistic action, the optimal policy, is 0.49% of the period's total GDP. The cost for inaction is 0.52%, while the cost for the optimal 2oC policy is 0.78% and the realistic, ambitious climate policy is 1.17%. For following centuries, the relative cost of inaction will increase.

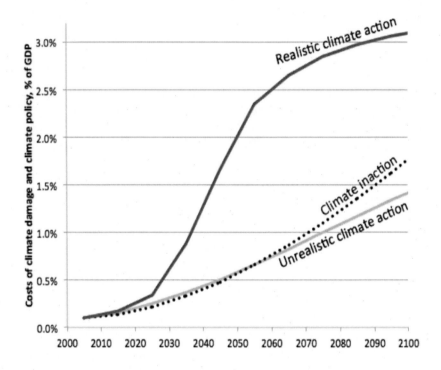

Figure 11. Total costs and benefits from inaction and action for the US. Black dotted line shows the cost of inaction. The light blue line shows the absolutely best-case cost of optimal, globally coordinated policies, with the cost of policy and the cost of residual climate damage. Dark blue line shows the more realistic cost of a mostly rich-country-led, ambitious, second-best climate policy along with residual climate damage. All calculations from RICE.

Two points are clear. First, global warming is by no means the most important part of the 21st century. Second, there is much greater scope for climate policies to make the total climate cost *greater* thought the 21st century.

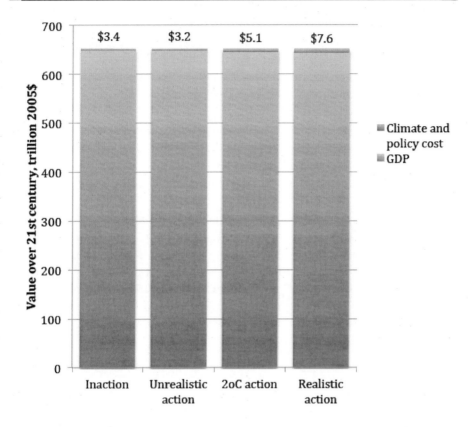

Figure 12. Costs of climate impacts and climate policy, and remaining GDP, for four different scenarios, over 21st century. The unrealistic action is the optimal action, generating a climate and policy cost of $3.2 trillion, and with a remaining GDP of $649.1 trillion. Realistic action is the mostlyrich-world scenario All calculations from RICE.

FAILED POLICIES TO TACKLE GLOBAL WARMING

This underscores the central question of how else to approach global warming.

The first realization needs to be that the current, old-fashioned approach to tackling global warming has failed. The current approach, which has been attempted for almost 20 years since the 1992 Earth Summit in Rio, is to agree on large carbon cuts in the immediate future. Only one real agreement, the Kyoto Protocol, has resulted from 20 years of attempts, with the 2009 Copenhagen meeting turning into a spectacular failure.

The **Kyoto approach is not working** for three reasons. **First**, cutting CO_2 is **costly**. We burn fossil fuels because they power almost everything we like about modern civilization. Cutting emissions in the absence of affordable, effective fossil fuel replacements means costlier power and lower growth rates. The only current, comprehensive global warming policy, the EU 20-20-20, will cost about $280bn/year.[15]

Second, the approach **won't solve the problem**. Even if everyone had implemented Kyoto, temperatures would have dropped by the end of the century by a miniscule 0.004_oC (0.007_oF). The EU policy will, across the century, cost about $20 trillion, yet will reduce temperatures by just 0.05_oC (0.1_oF).[16]

Third, **green energy is not ready** to take over from fossil fuels.[17] It is generally much costlier, its deployment does not in general create new jobs (because its higher, subsidized costs destroy jobs in the rest of the economy)[18], and because it typically produces electricity, which is not generated with oil, it doesn't reduce oil dependence[19]. Today, wind supplies 0.7% of global energy and solar about 0.1%, and even with very optimistic assumptions from the International Energy Agency, wind will supply only 2.4% in 2035 and solar 0.8%.[20]

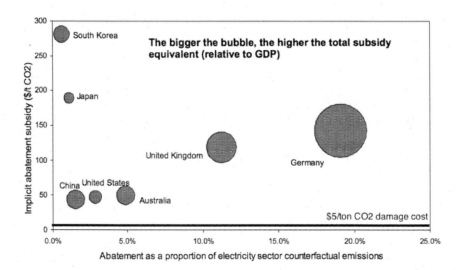

Figure 13. Abatement and implicit CO2 reduction cost for electricity, various nations. $5/ton CO2 damage insert for referece. In AUS$, which is almost equivalent to US$.[21]

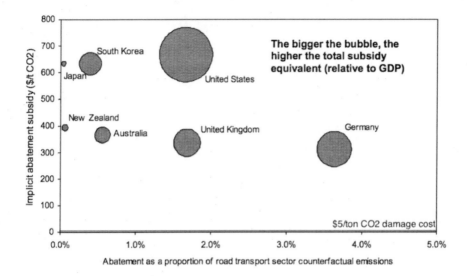

Figure 14. Abatement and implicit CO2 reduction cost for biofuels, various nations. $5/ton CO2 damage insert for referece. In AUS$, which is almost equivalent to US$.[22]

Because there is no good, cheap green energy, the almost universal political choices have been expensive policies that do very little. In Figure 13 we see how all major nations have managed to enact policies for electricity that cost a lot, yet do very little (Germany is leading the pack and still only reducing emissions from the power sector of 19% or 7% of the economy).

The cost per ton of CO2 avoided is universally far above the most likely $5/ton CO2 damage,[23] with China at the cheapest at 8 times the damage of at about $40, and South Korea at a phenomenal $280/ton CO2, 56 times higher than the damage cost. Germany pays each year about 0.3% of its GDP in electricity subsidies.

On biofuels, the excess cost is even more pronounced, and yet the emission reductions even smaller, as can be seen in Figure 14. Germany is paying 62 times too much or $310/ton CO2, reducing just 0.6% of its total emissions at a cost of $1.7bn. The US is paying a phenomenal 133 times too much, at $666/ton CO2, costing $17.5bn/year and reducing just 0.5% of its total emissions.

Yet, the cost is not just in economic terms. There is also increasing dissatisfaction with high energy costs in countries like the UK and Germany. In Germany the cost of electricity has risen 80% in real terms since 2000, as is evident in Figure 15. A fourth of all consumer energy costs are now direct subsidies to renewables.

Figure 15. Electricity price for households in Germany, 1978`2013.[24]

A BETTER POLICY APPROACH TO TACKLING GLOBAL WARMING

It is important to realize that the old-fashioned policies have failed. Current green technologies just won't make it[25]. The only way to move towards a longterm reduction in emissions is if green energy becomes much cheaper. If green energy was cheaper than fossil fuels, everyone would switch.

This requires breakthroughs in the current green technologies, which means focusing much more on innovating smarter, cheaper, more effective green energy.

Of course, pursuing an approach of R&D holds no guarantees—we might spend dramatic amounts on R&D and still come up empty in 40 years — but it has much higher likelihood of succeeding than our twenty-year futile attempts to cut carbon so far.

This was the recommendation of the Copenhagen Consensus on Climate, where a panel of economists including three Nobel laureates found that **the best long` term strategy** is to dramatically increase investment in green

R&D.[26] They suggested to 10-fold increase the current investment of $10bn to $100bn/year globally. This would be 0.2% of global GDP, and would entail a commitment of about $40bn from the US.

This approach would be significantly cheaper than the current policies (like the EU 20-20) and 500 times more effective. It is also much more likely to be acceptable to the developing countries.

The **metaphor** here is the **computer** in the 1950s. We did not obtain better computers by mass-producing them to get cheaper vacuum tubes. We did not provide heavy subsidies so that every Westerner could have one in their home in 1960. Nor did we tax alternatives like typewriters. The breakthroughs were achieved by a dramatic ramping up of R&D, leading to multiple innovations, which enabled companies like IBM and Apple to eventually produce computers that consumers wanted to buy.

This is what the US has done with fracking. The US has spent about $10bn in subsidies over the past three decades to get fracking innovation, which has opened up large new resources of previously inaccessible shale gas. Despite some legitimate concerns about safety, it is hard to overstate the overwhelming benefits. Fracking has caused gas prices to drop dramatically and changed the US electricity generation from 50% coal and 20% gas to about 40% coal and 30% gas.

This means that the US has reduced its annual CO_2 emissions by about 300Mt CO_2 in 2012.[27] This is about twice the *total* reduction over the past twenty years of the Kyoto Protocol from the rest of the world, including the European Union. At the same time, the EU climate policy will cost about $280 billion per year, whereas the US fracking is estimated to *increase* US GDP by $283 billion per year.

Conclusion

To answer the committee's question on the US cost of inaction and action for climate change, the short summary is this:

- The total, discounted cost of inaction on climate change over the next five centuries is about 1.2% of total discounted GDP.
- The cumulative cost of inaction towards the end of the century is about 1.8% of GDP
- While this is not trivial, it by no means supports the often apocalyptic conversation on climate change.

- The cost by the end of the century is equivalent to losing one year's growth, or a moderate, one-year recession.
- The cost of inaction by the end of the century is equivalent to an annual loss of GDP growth on the order of 0.02%.
- However, policy action as opposed to inaction, also has costs, and will still incur a significant part of the climate damage. Thus, with extremely unrealistically optimistic assumptions, it is possible that the total cost of climate policy action will be reduced *slightly* to 1.5% of GDP by the end of the century.
- It is more likely that the cost of climate action will end up costing upwards of twice as much as climate inaction in this century - a reasonable estimate would be 2.8% of GDP towards the end of the century.
- Thus, for the first half century, it is absolutely certain that any climate action will have greater total costs than inaction. For the second half of the century it is very likely that any climate action will have greater costs than climate inaction.
- While it is possible to design clever, well-coordinated, moderate climate policies that will do more good than they will cost, it is much more likely that the total costs of climate action will be much more expensive than climate inaction.
- To tackle global warming, it is much more important to dramatically increase funding for R&D of green energy to make future green energy much cheaper. This will make *everyone* switch when green is cheap enough, instead of focusing on inefficient subsidies and second-best policies that easily end up costing much more.
- It is likely that the percentage cost to the US budget is in the same order of magnitude as that of the percentage costs to the US economy.

End Notes

[1] Figure 1, p912, Richard S.J. Tol 2013: "Targets for global climate policy: An overview" in *Journal of Economic Dynamics & Control* 37 (2013) 911–928.

[2] Figure 4.1 in Gary W. Yohe, Richard S.J. Tol,, Richard G. Richels, Geoffrey J. Blanford 2009: The Challenge of Global Warming, in Lomborg, B 2009: *Global Crises, Global Solutions*, 2nd edition, Cambridge University Press. http://www.copenhagenconsensus.com/Files/Filer/CC08/Papers/0%20Challenge%20Papers/C P_GlobalWarmingCC08vol2.pdf

[3] Calculated from Nordhaus DICE model 2010, http://nordhaus.econ.yale.edu/RICEmodels.htm

[4] William D. Nordhaus 2010: "Economic aspects of global warming in a post- Copenhagen environment" in *Proceedings of the National Academy of Sciences*, 107:26, p11721–11726, doi: 10.1073/pnas.1005985107

[5] p19, http://ipccHwg2.gov/AR5/images

[6] Calculated from Nordhaus RICE model 2010, http://nordhaus.econ.yale.edu/RICEmodels.htm

[7] Nordhaus 2010, p4, recalculated to per ton CO_2 and CPI corrected to 2013.

[8] Richard Tol 2010, Carbon Dioxide Mitigation, in Lomborg 2010 *Smart Solutions to Climate Change*, Cambridge UK, Cambridge University Press.

[9] Nordhaus 2010, p4, recalculated to per ton CO_2 and CPI corrected to 2013.

[10] "Although unrealistic, this scenario provides an efficiency benchmark against which other policies can be measured."

[11] The so-called "Copenhagen Accord with only rich countries." I will here assume no trading between the blocks.

[12] http://cdiac.esd.ornl.gov/GCP/carbonbudget/2013/

[13] Christoph Bohringer etal. 2009: "EU climate policy up to 2020: An economic impact assessment" *Energy Economics* 31 (2009) S295-S305; Christoph Bohringer etal. 2009: "The EU 20/20/2020 targets: An overview of the EMF22 assessment" *Energy Economics* 31 (2009) S268- S273; Richard S.J. Tol 2012: "A cost–benefit analysis of the EU 20/20/2020 package." *Energy Policy* 49 (2012) 288–295,

[14] Christoph Bohringer etal. 2009: "The EU 20/20/2020 targets: An overview of the EMF22 assessment" *Energy Economics* 31 (2009) S268-S273.

[15] Richard S. J. Tol (2010) *The Costs and Benefits of EU Climate Policy for 2020*, Copenhagen Consensus Center.

[16] Tol (2010).

[17] Isabel Galiana and Christopher Green (2010) *TechnologyKLed Climate Policy*, in Smart Solutions to Climate Change; Comparing Costs and Benefits, Cambridge University Press.

[18] Gürcan Gülen (2011) *Defining, Measuring and Predicting Green Jobs*, Copenhagen Consensus Center.

[19] Research by climate economist Bohringer even shows that, fully implemented, the EU 20-20-20 plan does not boost energy security. See: Christoph Bohringer and Andreas Keller (2011) *Energy Security: An Impact Assessment of the EU Climate and Energy Package*, Copenhagen Consensus Center.

[20] International Energy Agency (2010) *World Energy Outlook 2000*, IEA/OECD.

[21] Pxxxvii, Australian Government Productivity Commission 2011: Carbon Emission Policies in Key Economies, http://www.pc.gov.au/projects/study/carbonHprices/report

[22] Pxxxix, Australian Government Productivity Commission 2011: Carbon Emission Policies in Key Economies, http://www.pc.gov.au/projects/study/carbon

[23] Richard S. J. Tol (2011). The Social Cost of Carbon, Annu. Rev. Resour. Econ. 2011. 3:419–43, doi: 10.1146/annurev-resource-083110-120028.

[24] Data from OECD (prices http://bit.ly/10IXX5J.

[25] For a sobering examination of the scale of the technological challenge, see: Isabel Galiana, Christopher Green (2009) *A Technology-led Climate Policy*, in Advice for Policymakers, Copenhagen Consensus Center. http://fixtheclimate.com/fileadmin/templates/page /scripts templavoila/COP15 Policy Advice.pdf

[26] Other influential research papers arguing for this approach include: Prins, Gwyn and Galiana, Isabel and Green, Christopher and Grundmann, Reiner and Korhola, Atte and Laird, Frank and Nordhaus, Ted and Pielke Jnr, Roger and Rayner, Steve and Sarewitz, Daniel and Shellenberger, Michael and Stehr, Nico and Tezuko, Hiroyuki (2010) *The Hartwell Paper:*

a new direction for climate policy after the crash of 2009. Institute for Science, Innovation & Society, University of Oxford; LSE Mackinder Programme, London School of Economics and Political Science; and also Steven F. Hayward, Mark Muro, Ted Nordhaus and Michael Shellenberger (2010) *Post-Partisan Power: How a limited and direct approach to energy innovation can deliver clean, cheap energy, economic productivity and national prosperity*. American Enterprise Institute, Brookings Institution, Breakthrough Institute.

[27] Zeke Hausfater 2013: Explaining and understanding declines in US CO_2 emissions.

In: Economic Costs of Inaction on Climate Change ISBN: 978-1-61728-031-3
Editor: Cheryl Griffin   © 2014 Nova Science Publishers, Inc.

Chapter 6

TESTIMONY OF W. DAVID MONTGOMERY, SENIOR VICE PRESIDENT, NERA ECONOMIC CONSULTING. HEARING ON "THE COSTS OF INACTION: THE ECONOMIC AND BUDGETARY CONSEQUENCES OF CLIMATE CHANGE"[*]

Chairman Murray, Ranking Member Sessions and Members of the Committee:

INTRODUCTION

I am honored by your invitation to testify on this very important topic. I am an economist and Senior Vice President at NERA Economic Consulting. I have worked on climate change issues and policy since 1988, when I was Assistant Director for Natural Resources and Commerce at the Congressional Budget Office and led CBO's study of the economic impacts of a carbon tax. Prior to that I was chief economist in the Office of Program Analysis and Evaluation in the Office of the Secretary of Defense and headed energy modeling and forecasting activities in the Energy Information Administration.

---

[*] This is an edited, reformatted and augmented version of testimony presented July 29, 2014 before the Senate Budget Committee.

After leaving government service, I continued to concentrate on climate policy for most of the last 25 years. I served as a Principal Lead Author of the IPCC's Second Assessment Report and as a Peer Reviewer of subsequent reports including the most recent. I have led numerous studies in which I and my colleagues at NERA used our economic models to estimate the costs and emission reductions of proposed climate policies including those included in the President's Climate Action Plan. I have published many articles on these topics in refereed professional journals, and had the privilege of contributing a chapter on black carbon mitigation to a volume on climate change edited by Professor Lomborg a few years ago.[1]

In the past few years I have taken a particular interest in the relative merits of mitigation and adaptation as responses to climate change risks, and in particular in the role of political and economic freedom in making it possible for poor countries to grow economically and at the same time to reduce their carbon intensity and become more resilient in adapting to climate change.

I am testifying on my own behalf today, and statements in this testimony represent my own opinions and conclusions and do not necessarily represent opinions of any other consultant at NERA or any of its clients.

SUMMARY

Today's hearing centers on the potential damage that climate change could cause and how that possible damage could affect the economy and the Federal budget. This is a very broad topic, and the questions that it raises cover nearly every aspect of our knowledge about climate change:

- How imminent and likely is damage from global warming?
- What is the government's role in reducing the potential damage from climate change?
- How much should be spent on public investments to "protect people" from climate risks?
- How much damage can be avoided by reducing greenhouse gas emissions?
- What will it cost the Federal budget and the U.S. economy to reduce emissions?
- How confident can we be that spending more now will reduce the likelihood or magnitude of future costs enough to justify the expenditure?

It is far from clear that recent weather events are anything more than normal variability in storm frequency and intensity, and the nature, timing and extent of damage from climate change remains highly uncertain. This does not imply that no action is justified, but it does imply that costs and avoided risks must be balanced carefully.

Unlike reducing greenhouse gas emissions, for which there are not adequate private incentives in the absence of government policies, there are quite sufficient incentives for private households and businesses to pay attention to risks of climate change. The role of government should be limited to revising priorities for public investments in light of climate risks, and reforming existing policies and programs that distort incentives for risk minimization, such as subsidized flood insurance.

Since the public investments that could be justified as a defense against climate change will be under the jurisdiction of the same Congressional committees and executive agencies now dealing with similar activities, there will be natural incentives in Congress and the agencies to propose increases in spending beyond what a critical evaluation of costs and risk reduction would justify. Expanding the role of government into activities that could be done perfectly well by the private sector will not save budgetary resources, nor will overinvestment in areas where government does have a responsibility. There are also questions of timing. At a time when we face threats around the world, national security might be better served by reversing planned reductions in military manpower and force structure than by increasing funding for climate related activities in DOD. Thus critical evaluation of such budget proposals will be very important.

Policies to reduce greenhouse gas emissions, such as the Administration's Climate Action Plan (CAP), have also been rationalized on the grounds that spending now will avoid higher costs later. I have used our Ne$_w$ERA model of the U.S. economy to estimate the economic cost and budgetary impact of policies sufficient to achieve the CAP goal of reducing emissions to 17% below 2005 levels by 2020. The regulatory policies favored by the Administration would be likely to reduce the average household's disposable income by about $1000 in 2020, reduce Federal revenues by over $150 billion due to reduced economic growth, and cause electricity prices to rise by about 7%.

Holding U.S. emissions at 17% below 2005 levels all the way to 2040 would reduce cumulative global emissions over that period by less than 2%, because of the declining share of the U.S. in global emissions. That would take as little as three-hundredths of a degree and no more than one ten-tenth of a degree off the rise in global average temperatures that might occur otherwise. Damages to the U.S. would probably be reduced by about the same 2 percent.

This leaves the question of whether there is a national security interest in climate change due to its likelihood of increasing conflicts or effects on U.S readiness. It is true that most of the damage from climate change will not occur in the U.S. but rather in poor countries in equatorial regions -- in other words, in regions where failed states, rapacious dictators, and ethnic and religious violence are endemic. The paltry difference in global warming that the US can make by reducing emissions will not help those countries. I believe that we have both a national interest and moral obligation to provide effective, community based aid to those countries to assist them in adaptation. The overwhelming evidence, however, is that resilience to climate change -- that is, ability to adapt -- is greatest in countries whose open political systems and free market economies provide both the incentives and the stability for private investments in adaptation, and impossible to achieve in others. Thus it would be far better to concentrate on ways to bring about open political and economic systems in these poor countries than to engage in more of the planned, top down aid that has failed to alleviate poverty or violence up to now. Absent such changes, providing adequate budgets for national defense to deal with threats from those regions will remain the same high priority no matter how they are affected by climate change.

BUDGET IMPACTS COME FROM POLICY CHOICES

Climate analysts use the word "mitigation" to describe actions intended to reduce greenhouse gas emissions or their effects on global temperature, and "adaptation" to describe human responses that can lessen the damage from higher temperatures. It is convenient to put policy choices into one or the other of these categories. Mitigation policies are intended to avoid future damage from climate change by reducing greenhouse gas emissions and limiting the range of likely increases in global average temperatures. Adaptation policies are intended to reduce the damage from climate change if and when it does occur. Both types of policies can have effects on the economy and the budget, but they differ greatly in their cost-effectiveness in reducing risks.

Possible climate impacts form the basis for either mitigation or adaptation. For example, the President's Climate Action Plan states that " we are already feeling [climate] impacts across the country and the world." In this, the President goes well beyond what the IPCC has stated. We have indeed experienced weather events that might in the future be made worse by rising global temperatures, but the evidence that any recent events are caused by global warming is not convincing even to the IPCC.[2] The events are well within normal variability of the weather system, do not need a driver of rising global temperature to happen. Most of record damage due to storms is clearly attributable to greater development in areas known to be vulnerable and not to an increase of the hazard. Fixing the incentives to locate in locations at risk is far more cost-effective than encouraging and then protecting unwise investments through mitigation or adaptation.

Although it is true that demanding certainty before acting is rarely a good risk management strategy, always assuming the worst and acting as if it is sure to happen without immediate action is equally bad risk management. So is insisting on doing something even though it is too late or too little to matter.

A prudent balancing of costs and risks is necessary, and that is very hard to do given the present lack of quantification and high uncertainty about what the effects of climate change will be. The range of temperature increases predicted as a result of a doubling of greenhouse gas concentrations is wider in the most recent IPCC Fifth Assessment Report than it was in the Fourth, from 1.5 to 4.5 degrees Celsius, and global temperature increase has stalled for the past 15 years. If the cause is that uncertain, the effects cannot be any less uncertain. Although studies of the potential damages of events hypothesized to be caused by climate change, known as "effects research," have proliferated, integrated assessment modelers have not yet succeeded in extending their models that predict temperature change to generate estimates of the effects of temperature increase and the damages that they would cause.[3] Moreover, the effects of temperature increase are likely to be so localized and model results are so inconsistent about global effects that global or national planning is most likely to do the wrong thing in the wrong place.[4] Reduction of greenhouse gas emissions from the United States faces the high likelihood that the countries that will emit the most emissions over the next decade, including China, India and Russia, will do nothing to reduce their emissions, leaving climate risks about the same no matter what the U.S. does.

MITIGATION POLICY

Thus the first questions about mitigation policy have to be about its present economic costs and budgetary effects and its possible future benefits. Implementation of the President's Climate Action Plan in particular would have significant effects on revenues and outlays as well as negative impacts on the economy as a whole.

Economic and Revenue Impacts of Mitigation Policy

In order to assess the consequences of the Climate Action Plan, my colleagues and I used NERA's Ne$_w$ERA model of the U.S. economy. Ne$_w$ERA is a computable general equilibrium model of the U.S. economy that has been used by a wide range of clients for assessments of energy and environmental policies, including the U.S. Department of Energy in its evaluation of the public interest in allowing natural gas exports.[5] For this study, we used a version of the model that has a baseline for taxes and expenditures based on CBO's long term budget outlook and a detailed representation of income taxes and drivers of spending.

Economic Impacts

We took the Climate Action Plan goal of a 17% reduction in greenhouse gas emissions below 2005 level by 2020, and assumed the same cap would remain in place thereafter. We found that is the goal were achieved in the most cost-effective possible way, by achieving an equal marginal abatement cost across all possible ways of reducing emissions, there would still be impacts on energy prices, GDP, and federal, state and local tax revenues:

- Energy prices: 7% higher residential electricity prices in 2020, 23 cent per gallon or higher gasoline prices, and about a 10 cent per million BTU increase in natural gas prices due to increased demand for natural gas for power generation.
- Real disposable income: Less by over $200 per household in 2020
- Tax revenue: Federal revenue down by $40 billion in 2020 and State and local revenue by $4 billion.

Energy prices occur because limiting greenhouse gas emissions requires moving to higher cost sources of energy, and abandoning capital investments that rely on coal or oil to replace them with more costly sources of energy. These costs are all passed on to consumers in the form of higher prices. These cost increases and the demand for resources to replace existing capital prematurely ripple through the economy and reduce wages, returns to capital and GDP. Higher prices and lower wages and returns on investment lead to lower disposable income for households and to a shrinking of the tax base that reduces Federal and state revenues.

Costs would be higher with actual policies that leave some sectors out and drive others too far. The CAP is not a broad and uniform policy that puts a uniform price on carbon dioxide emissions wherever they occur in the economy, that would concentrate effort on reducing emissions where it is most cost-effective. Instead, the Climate Action Plan lists a series of regulatory measures to force electric utilities, consumers and motorists to switch fuels and use less energy. The stated components of the CAP include

- EPA's CO2 emission standards for new and existing powerplants
- Tightening new car fuel economy standards
- More strict appliance efficiency standards

It is dubious that these specific measures alone[6] could achieve the stated goal, but they do identify a regulatory approach to climate policy that would be much more costly than the estimates I gave for a minimum cost approach if it were expanded sufficiently to achieve the 17% reduction. In a paper published last year in the Energy Journal, my colleagues compared the cost a policy that achieved emission reductions at least cost to the cost of various regulatory policies that achieved the same goal.[7]

The key figure from their paper is reproduced below. The horizontal axis measures cumulative emission reductions from 2010 - 2050 in millions of metric tons, and the vertical axis measures costs in net present value over the same time period, in trillions of dollars. The curved line represents the minimum cost, with ideal policies, at which emission reductions could achieved. Any point below the line represents a policy that achieves the same emission reduction at higher cost. The policies labeled TRN includes transportation measures such as fuel economy standards and renewable fuel standards. The policy labeled CES is a policy that requires utilities to source progressively larger percentages of their generation from "clean" sources including natural gas and renewables. The point labeled RPS is a more conventional renewable portfolio standard for electricity generators. The policy labeled REG contains all of the above plus tightened energy efficiency standards. Thus REG contains the same regulatory elements as the Climate Action Plan, and applies them to the electricity generation sector, transportation sector, and household sector with sufficient severity to lead to cumulative emission reductions of about 30 million metric tons between 2010 and 2050.

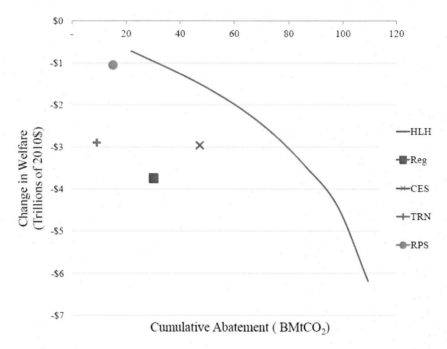

Changes in Discounted PV of Welfare from 2010-2050 for Regulatory Mandates Compared to Efficient Frontier (Trillions of 2010$).

If the goal of the Climate Action Plan is to reduce emissions to 17% below 2005 levels by 2020 were met, and emissions were held at this level from 2020 to 2050, we estimate that cumulative emissions would be reduced by about the same 30 million metric tonnes as the REG policy in our study. Thus the REG policy in the figure above gives a good indicator of what the cost would be if the Administration's regulatory approach were made sufficiently stringent to actually achieve its stated goal.

The picture reveals that taking a regulatory approach, with CO2 emission regulations, requirements for generation of electricity from "renewable" sources, new car fuel economy standards and "renewable" fuel standards together with even tighter efficiency standards on appliances and other consumer durables would cost about 4 times as much as a least cost policy.

That implies a cost by 2020 of about $1000 per household and, if budgetary impacts of the regulatory policy are proportional to its other impacts, a loss of over $150 billion in revenue in FY 2020.

Revenue Impacts

Impacts on GDP and personal income translate into lower tax revenues. Our analysis has shown that even an ideal carbon tax would have to devote up to 40% of its revenues to make up for lost revenues elsewhere in the economy due to drag on the economy. Regulatory measures that provide no revenue offset and lead to greater losses in GDP would have even larger effects on revenues.[8]

Thus the mitigation policy approach of this Administration will unambiguously reduce revenues, probably on the order of $150 billion by FY2020.

OUTLAYS FOR MITIGATION

These would not be the only effects of mitigation policies on the budget. There are many proposals mentioned in the CAP and proposed policies that also increase the budget deficit from the outlay side. These include:

- Extended tax preferences for solar, wind, and other renewables. These subsidies hide the higher cost of renewables relative to fossil fuels and shift both that cost and windfalls to economic renewables onto the taxpayer. But those costs do not go away. The impacts of using such measures to achieve CAP goals were not included in my estimates of lost revenue, and would make revenue losses even larger if they were extended.
- Loan guarantees are likely to have adverse consequences for the budget as well. They contain a built-in bias toward failure. Providing an upfront credit subsidy will not keep a project in business if it cannot cover its operating costs, as recent failures in battery and solar technology prove. Even the requirement for a loan guarantee fee to cover expected losses leads to adverse selection, because the fee those with a worse than average probability of default will be most willing to pay it, and companies with better than average likelihood of success will not.[9]
- Demonstration projects are at the wrong end of the RDD&C spectrum for government to be involved. The appropriability of R&D increases as it comes closer to being commercial, and the need for government involvement disappears. Demonstration projects in practice have led

to diversion of R&D funds from more fundamental research that could lead to real breakthroughs and cost reduction, and do not lead to adoption of the demonstrated technology when no long term incentives for replacing fossil fuels like carbon taxes are in place.[10]

These budgetary and economic impacts of tax subsidies and loan guarantees would increase the loss in revenue above $150 billion in 2020, and are additional to the revenue losses due to regulatory programs that divert productive investment and put a brake on economic growth.

POSSIBLE AVOIDED DAMAGES

There is very little policy or budget guidance to be found in discussions of the costs of unchecked climate change. The better question is what damage could be avoided if specific goals were achieved, and what the cost would be of policies that could realistically be expected to reach those goals.

This is pretty much basic environmental economics. The next point is also pretty uncontroversial, that the avoided damages are much more uncertain than the costs. We can and do employ scenario analysis to provide an understanding of how the cost of achieving a specific reduction in emissions depends on uncertainties about future technology developments and about how much it will cost households and business to change behavior and investments in buildings and equipment. But for avoided damage the uncertainties are much wider.

It is, however, possible within these ranges to distinguish the difference between very big and very little. EIA projects that cumulative US emissions between 2015 and 2040 will be approximately 14% of cumulative global emissions. Using EIA's most recent projection of BAU US emissions, the goal of the Administrations Climate Action Plan to reduce emissions to 17% below 2005 by 2020 and assuming they are kept at that level would reduce global cumulative emissions from 2015 to 2040 by less than 2%. Thus whatever the range of global temperature increase is projected to be between now and 2040, the CAP would reduce that increase by less than 2%, and therefore would likely reduce avoided damages by a similar percentage. Thus if the range is 1.5 to 4.5 degrees C, the effect of the CAP would be to change the range to 1.5 to 4.4 degrees, if we stick to rounding to one decimal place. There is no climate model that can tell the difference in effects between those two ranges.

Thus mitigation in the U.S. alone is not likely to reduce U.S. contingent damages by as much as the policies cost, especially if stringent regulatory policies are adopted.

ADAPTATION POLICY

If we accept that the Administration's policies will not affect damages to the U.S. or the rest of the world, adaptation becomes a high priority. In addition to its goal for reducing greenhouse gas emissions, the CAP states that "...we must also prepare for the impacts of a changing climate that are already being felt across the country [sic]. Moving forward, the Obama Administration will help state and local governments strengthen our roads, bridges, and shorelines so we can better protect people's homes, businesses and way of life from severe weather."

It is good to focus on adaptation. The U.S. can be very resilient if we return to principles of free markets and private initiative. That is why most studies conclude that most of the damages from climate change will occur in poor equatorial countries. Most of the benefits of global reductions in greenhouse gas emissions would go to those countries. That is not a bad thing, but there are much more direct and potentially cost-effective ways to help those countries than costly and largely ineffective efforts to reduce emissions. Effective aid for local adaptation is one. For the U.S., all we really need to do is avoid damaging our built-in resilience through badly designed policies. As my friend and colleague put it in Forbes recently "... the main U.S. line of defense against the risks of climate change ... remains a free and productive economy."[11]

ECONOMIC ISSUES IN DESIGNING ADAPTATION STRATEGIES

Economists who have studied climate change generally agree that rational adaptation can substantially reduce the potential damage from climate change,[12] and that in an institutional setting that does not distort the natural economic incentives to avoid risk, the private sector is quite capable of adopting many appropriate responses on its own. There are also public goods involved in adaptation, including the classic public goods of R&D, public health, roads, dams and flood protection. Resilience toward climate change is also a function of how well a system performs at providing public goods. Thus to me there are three fundamental requirements for effective adaptation policy in the United States:

- Understanding what types of adaptations should be left to the private sector and which are the responsibility of government. The criterion should not be "people's homes need protecting" but "there are systematic public goods involved that justify public investment rather than relying on the clear private incentive to manage risks to ones own property."
- Maintaining the economic freedom and property rights that create appropriate incentives for private investment to avoid risks of climate change. Unlike many countries of the world, our system of private property and free enterprise provides a framework for appropriate incentives and has led to successful adaptation to all sorts of changes affecting the economy. But these incentives can be diluted or distorted by government programs that provide free insurance before or after the fact or otherwise subsidize development in vulnerable areas.
- Limiting public policy toward adaptation to a. elimination of subsidies and other distortions that reduce private adaptation incentives by creating moral hazards b. investments in true public goods that have an acceptable balance of cost and climate risk reduction.

Poor countries face much greater challenges than these in achieving any kind of adaptation. Where our problem is adapting sensibly and cost-effectively, their problem is adapting at all. Many studies have shown why it is that poor countries, especially in equatorial regions where the potential effects of climate change would be the largest, are not likely to be able to adapt effectively. These include violence and insecurity that makes any investment questionable, rulers who keep their people in poverty while appropriating any economic surplus or foreign aid for their own benefit, and lack of secure property rights and land tenure that are fundamental to incentives to invest.[13] They also include closed political systems that exclude most of their population from meaningful participation and carry out public works projects to benefit their narrow base of supporters and not the country as a whole.

Countries like Botswana that have achieved free market and political systems have already been successful in mitigating the risk of weather events and instability of agricultural prices, and many Central African countries will remain poor and vulnerable as long as violence is a more attractive option than participation in the political system.[14] I am firmly convinced that moving a country from a political order like that in, for example, the Sudan, to a political order like that in Botswana would improve its standard of living and reduce the potential for conflict and damage from climate change more than would any conceivable action to reduce global greenhouse emissions.

POTENTIAL PITFALLS IN ADAPTATION POLICY

Nor is the United States immune to distorted incentives and government policies that frustrate or misdirect adaptation. Our current policies already distort incentives in a way that increases vulnerability to extreme weather events and inflates estimates of the need for public investment to protect socially unwise private investments. The principle one is subsidized flood insurance, that encourages people to build in areas known to be vulnerable. A more hidden incentive is provided by Federal funding for reconstruction after a disaster hits; although solidarity with those who have been harmed justifies aid, providing the aid by rebuilding the areas that were damaged just reinforces the incentive to downplay risks. Most of record damage due to storms is clearly attributable to greater development in areas known to be vulnerable and not to an increase of the hazard. Fixing the incentives to locate in locations at risk is far more cost-effective than encouraging and then protecting unwise investments. Agricultural disaster assistance can have the same effect. The moral hazard that future policies could create must be looked at carefully if private adaptation is to play the full role that it can.

In terms of the design of public investment programs, I see three counterproductive dynamics at work, that if left unchecked are likely to greatly increase budgetary demands and reduce the effectiveness of adaptation measures. They are:

- Scientifically unjustified attribution of current weather events to climate change
- Using adaptation as a convenient rationale for pork barrel projects
- Making climate change an excuse for extension of agency missions and larger budgets

The first of these is a simple error, though it many cases it is indulged in by those who do know better.[15] The other two are consequences of a dysfunctional system in which policies are pursued for the benefit of incumbents and their constituencies rather than for broader national objectives.[16]

Even in cases when certain activities are clearly the responsibility of government, distinguishing the wheat from the chaff in proposed investments in adaptation is more difficult than it might appear. Not one of potential public investments in adaptation is unique to climate change. Public health, public buildings, roads, dams, levees, fire and flood protection have well organized constituencies and agencies that promote, build and oversee them. These are also (with the exception of public health) the areas in which pork barrel politics was invented. Thus the natural Congressional and bureaucratic incentives line up to encourage unnecessary spending on adaptation, and a critical attitude toward any such plans is warranted.

The Budget Committee has always tried to resist these tendencies. Two things that the Committee can do in the case of adaptation is to consider the proper role for government and scrutinize specific funding requests to ensure they represent cost-effective solutions to problems within government role.

Although it is true that demanding certainty before acting is rarely a good risk management strategy, always assuming the worst and acting as if it is sure to happen without immediate action is equally bad risk management. So is insisting on doing something even though it is too late or too little to matter.

A prudent balancing of costs and risks is necessary, and that is very hard to do given the present lack of quantification and high uncertainty about what the effects of climate change will be. The range of temperature increases predicted as a result of a doubling of greenhouse gas concentrations is as wide in the most recent IPCC Fifth Assessment Report as it was in the first. If the cause is that uncertain, the effects cannot be any less uncertain. Although studies of the potential damages of events hypothesized to be caused by climate change, known as "effects research," have proliferated, integrated assessment modelers have not yet succeeded in extending their models that predict temperature change to generate estimates of the effects of temperature increase and the damages that they would cause. Moreover, the effects of temperature increase are likely to be so localized and model results are so inconsistent about global effects that global or national planning is most likely to do the wrong thing in the wrong place.[17]

WHERE ADAPTATION IS MOST NECESSARY

Despite all this, I agree that "To lower our national security risks, the United States should take a global leadership role in preparing for the projected impacts of climate change."[18] But I recommend a very specific type of response. because I am convinced that most assessments of what can be done are so blinded by political correctness and diplomacy that they will not properly attribute the cause of vulnerability to failed states, rapacious ruling elites, and systems that fail to provide either economic or political freedom. They also continue the error of recommending top down planned solutions rather than recognizing that effective adaptation, like effective poverty reduction and wildlife conservation, must occur at the community level.[19]

In the past decade, Botswana has experienced a surge of economic growth and reduction in poverty, as well as implementing systems that have substantially reduced risks of drought and price fluctuations for the agricultural sector. At the same time, Zimbabwe has continued its process of expropriation of white farmers and assignment of those lands to cronies of dictator Mugabe, with the result that agricultural production has collapsed, poverty and hunger have increased, and vulnerability to climate change greatly increased.

Regimes reap the harvest from any the seeds of conflict that might be planted by adverse environmental conditions, and conditions that may lead to conflict in a closed political and economic society are much less likely to in an open society. Indeed, discovery of sufficient wealth in a country to make fighting over who will control it has triggered conflict where poverty was long tolerated. Nor is environmental degradation new as a cause of conflict. Before their war with white settlers, the cattle-raising Zulu warriors moved south into lands settled by other tribes, took them over and slaughtered the population to provide room for their herds as they depleted northern grazing lands. These conditions may be made worse by climate change, but the small difference that unilateral U.S. action can make to global warming in the current international setting will have no noticeable effect on the risks. To the extent that these conflicts affect U.S. national interests, a much wiser investment would be in a sufficiently strong military to deal with threats to us and humanitarian interventions around the world.

If we really want to help globally, there is clear evidence that most can be accomplished through effective support at a community level for locally-designed and implemented adaptation measures in Africa and poor Asian countries where the real vulnerability exists, not nugatory mitigation that helps no one.

This concludes my prepared testimony and I look forward to your questions.

End Notes

[1] "Black Carbon Mitigation." With R. Baron and S. Tuladhar. Chapter 4 in Smart Solutions to Climate Change – Comparing Costs and Benefits, Bjorn Lomborg (ed.), Cambridge University Press, 2010.

[2] For example, in its Fifth Assessment Report (FAR) the IPCC states that " Economic losses due to extreme weather events have increased globally, mostly due to increase in wealth and exposure, with a possible influence of climate change (low confidence in attribution to climate change)."

[3] Again the FAR states "In recent decades, climate change has likely contributed to human ill-health although the present world-wide burden of ill-health from climate change is relatively small compared with effects of other stressors and is not well quantified."

[4] Robert Mendelsohn, op cit.

[5] For a description of the model and how we represent policies like the CAP in it, see Sugandha D. Tuladhar, Sebastian Mankowski, and Paul Bernstein. The Interaction Effects of Market-Based and Command-and-Control Policies. Energy Journal, Vol. 35, No. SI1.

[6] THE PRESIDENT'S CLIMATE ACTION PLAN Executive Office of the President June 2013. pages 6 - 9.

[7] Tuladhar et. al.

[8] This is a standard finding in the literature on the "double dividend" literature, see Lawrence H. Goulder, "Environmental taxation and the double dividend: A reader's guide" International Tax and Public Finance August 1995, Volume 2, Issue 2, pp 157-183

[9] See my chapter in Pure Risk: Federal Clean Energy Loan Guarantees, Nonproliferation Policy Education Center Apr 2012.

[10] See Kenneth J. Arrow, Linda Cohen, Paul A. David, Robert W. Hahn, Charles D. Kolstad, Lee Lane, W. David Montgomery, Richard R. Nelson, Roger G. Noll and Anne E. Smith. "A Statement on the Appropriate Role for Research and Development in Climate Policy" Economists Voice, February 2009. Lee Lane, W. D. Montgomery and A. Smith, "R&D Policy." in A Taxing Debate: Climate Policy Beyond Copenhagen. Growth No. 61, Committee for Economic Development of Australia, August 2009.

[11] http://www.forbes.com/sites/realspin/2014/07/17/the-risky-business-of-a-carbon

[12] Robert Mendelsohn, "The Impact of Climate Change on Land," with commentary by W. David Montgomery in Climate Change and Land Policies, edited by Gregory K. Ingram and Yu-Hung Hong, Lincoln Institute for Land Policy, April 2011

[13] See, for example, Paul Collier, The Bottom Billion: Why the Poorest Countries are Failing and What Can Be Done About It and William Easterly The Tyranny of Experts: Economists, Dictators, and the Forgotten Rights of the Poor.

[14] Robert Bates, When Things Fell Apart: State Failure in Late-Century Africa (Cambridge Studies in Comparative Politics) 2008.

[15] For example, the late Stephen Schneider characterized some climate scientists as taking a "sound-byte" approach that he found reprehensible but understandable: "And like most people we'd like to see the world a better place, which in this context translates into our working to reduce the risk of potentially disastrous climatic change. To do that we need to get some broad based support, to capture the public's imagination. That, of course, entails getting loads of media coverage. So we have to offer up scary scenarios, make simplified, dramatic statements, and make little mention of any doubts we might have." See American Physical Society, APS News, August/September 1996, p. 5. Nevertheless, the practice has continued.

[16] See Morris Fiorina, Congress: Keystone of the Washington Establishment, Revised Edition, 1989 and Lee Lane and W. David Montgomery, "Political Institutions and Greenhouse Gas Controls." AEI Center for Regulatory and Market Studies, Related Publication 08-09. Revised August 2010.

[17] Montgomery, Lincoln Institute, op cit.

[18] National Security and the Accelerating Risks of Climate Change May 2014 CNA Military Advisory Board, p. 5

[19] Easterly, op cit.

INDEX

#

21st century, 45, 83, 86, 87, 88, 89

A

abatement, 83, 102
abuse, 60
access, 49, 57, 59, 60, 61, 67
acidic, 4, 7, 50
acidity, 22
adaptation, 11, 49, 53, 54, 56, 57, 58, 66, 68, 72, 98, 100, 107, 108, 109, 110, 111
adaptations, 108
adjustment, 9
adverse event, 50
Africa, 22, 65, 66, 111, 112
age, 10
agencies, 48, 51, 55, 56, 58, 59, 60, 69, 99, 110
agricultural producers, 72
agricultural sector, 111
agriculture, 11, 35, 42, 43, 59
Alaska, vii, 1, 52, 53
amphibia, 22
appropriations, 40
aquifers, 66
Arctic sea ice, vii, 3, 7, 20, 22
Asia, 65
Asian countries, 112

assessment, 11, 16, 33, 64, 65, 84, 86, 95, 101, 110
assessment models, 11, 33
assets, 39, 49, 54, 70
atmosphere, 2, 3, 4, 12, 13, 22, 35, 50
attribution, 109, 112
audit, 51
authorities, 69
authority, 50, 60
awareness, 68

B

Bangladesh, 66
base, 48, 52, 53, 69, 70, 78, 79, 108
basic research, 8
beef, 42
benchmarking, 45
benefits, viii, 2, 3, 4, 6, 8, 11, 26, 35, 48, 50, 56, 58, 81, 82, 83, 84, 86, 87, 88, 93, 101, 107
bias, 105
biodiversity, 10, 22
biological systems, 22, 26
blueprint, 68
Botswana, 45, 108, 111
BTU, 102
Budget Committee, 39, 47, 63, 75, 97, 110
budget deficit, 105
budgetary resources, 99
burn, 45, 55, 90

Bush, President George W., 71
businesses, 3, 6, 40, 44, 99, 107

C

CAP, 99, 103, 105, 106, 107, 112
carbon, vii, 1, 3, 4, 6, 8, 12, 13, 16, 22, 26, 32, 33, 34, 35, 36, 43, 44, 45, 50, 68, 81, 83, 86, 89, 92, 95, 97, 98, 103, 105, 106, 112
carbon dioxide, vii, 1, 3, 22, 50, 103
carbon emissions, 4, 13, 26, 36
carbon-based energy, vii, 1
case study, 68
catalyst, 64, 66, 67, 73
catastrophes, 3, 4, 7, 21, 24, 26
cattle, 42, 111
challenges, 25, 37, 39, 57, 65, 67, 68, 69, 71, 108
children, 45
China, 65, 78, 82, 83, 84, 85, 91, 101
circulation, 7
cities, 41, 66
City, 41
civilization, 90
Clean Air Act, 31, 35
clean energy, 39, 44, 45
clients, 98, 102
climate change issues, 97
climates, 35
closure, 57
CO2, vii, 1, 2, 3, 4, 6, 7, 9, 13, 15, 16, 18, 26, 27, 28, 34, 36, 44, 81, 83, 85, 90, 91, 93, 95, 96, 103, 104
CO2 emissions, vii, 1, 2, 3, 4, 6, 44, 85, 93, 96
coal, 13, 35, 43, 44, 93, 102
coastal region, 70
coffee, 40
Cold War, 67
collaboration, 61
commercial, 13, 69, 105
communities, 40, 42, 59, 68, 69, 72
community, 25, 65, 66, 69, 100, 111
complexity, 53
computer, 93
conditioning, 72
conflict, 10, 64, 66, 73, 109, 111
Congress, 50, 60, 99, 113
Congressional Budget Office, 57, 61, 97
consensus, vi, vii, viii, 1, 6, 35, 64, 75, 92, 95
construction, 52, 56
consumers, 72, 93, 102, 103
consumption, 16, 24, 25, 36, 78
contingency, 64
control measures, 53
cooling, 11, 71
cooperation, 6, 67, 73
coordination, 13, 15
correlation, 36
cotton, 39, 72
CPI, 95
crises, 69
critical infrastructure, 48, 51, 68, 70, 71
crop(s), 40, 41, 42, 72
customers, 45

D

damages, 2, 5, 9, 11, 35, 37, 48, 49, 50, 78, 79, 80, 82, 83, 84, 86, 87, 101, 106, 107, 110
data set, 16
decision makers, 51, 56, 57, 59
Delta, 66
Department of Defense, 48, 51
Department of Energy, 13, 68, 102
Department of Homeland Security, 57, 61, 71
dependent variable, 18, 20, 36
deposits, 22
destruction, 22
developed countries, 16, 33, 34
developed nations, 19
developing countries, 6, 15, 34, 84, 93
diplomacy, 70, 111
directives, 71
disaster, 40, 42, 43, 50, 57, 66, 67, 69, 109
disaster assistance, 40, 43, 50, 57, 109

disaster relief, 40, 41, 42, 43, 69
dispersion, 67
displacement, 10, 65
disposable income, 99, 102, 103
dissatisfaction, 91
distortions, 108
distribution, 21
diversification, 37
diversity, 3, 39
drought, 39, 41, 42, 43, 52, 53, 55, 65, 69, 71, 111
droughts, vii, 3, 42, 47, 49, 50, 64

E

Earth Summit, 89
eBay, 39
economic activity, 10, 11
economic consequences, viii, 2, 3, 25, 26, 33
economic damage(s), vii, 1, 2, 3, 4, 5, 9, 11, 12, 25
economic growth, 25, 99, 106, 111
economic growth rate, 25
economic incentives, 107
economic losses, 24
economic systems, 7, 100
economics, 2, 4, 21, 24
ecosystem, vii, 3, 7, 9, 22
Egypt, 66
electricity, 35, 40, 45, 71, 72, 90, 91, 93, 99, 102, 103, 104
emergency, 70
emission, vii, 1, 8, 13, 91, 98, 103, 104
emitters, 84, 85
empirical studies, 10
employment, 16
energy, vii, 1, 3, 13, 14, 43, 44, 45, 65, 66, 67, 71, 76, 81, 90, 91, 92, 94, 95, 96, 97, 102, 103
energy efficiency, 45, 103
energy prices, 102
energy security, 3, 95
engineering, 57
environment(s), 35, 45, 67, 69, 70, 95

environmental conditions, 111
environmental degradation, 111
environmental economics, 106
Environmental Protection Agency, 4, 32
EPA, 4, 35, 103
equilibrium, 7, 21, 25, 26, 35, 78, 102
equipment, 67, 70, 106
ERA, vi, 97
erosion, 10
EU, 78, 84, 85, 86, 90, 93, 95
Europe, 34, 65
European Union, 16, 93
evacuation, 65, 69
evaporation, 67
Everglades, 66
everyday life, 71
evidence, 10, 13, 20, 22, 35, 44, 51, 100, 101, 111
Executive Order, 59, 61
executive orders, 56
expenditures, 4, 102
experimental design, 16
exports, 102
exposure, 40, 41, 44, 51, 53, 55, 57, 59, 112
extinction, vii, 3, 10, 21
extraction, 66, 67
extreme precipitation, 53
extreme weather events, 10, 40, 42, 43, 47, 48, 49, 51, 52, 53, 57, 59, 65, 69, 70, 109, 112

F

failed states, 100, 111
farmers, 41, 72, 111
farms, 66
federal facilities, 49, 52
federal government, 41, 43, 48, 49, 51, 54, 55, 57, 58, 59, 60, 70, 79
Federal Government, 48, 50
FEMA, 41
fencing, 49, 53
financial, 3, 24, 40, 44, 50
fish, 55
flooding, 10, 40, 41, 48, 50, 53, 68, 71

floods, 41, 47, 50, 64
fluctuations, 111
food, 22, 40, 42, 65, 66
food production, 66
force, 44, 69, 99, 103
Ford, 28, 33
forecasting, 97
foreign aid, 108
fraud, 60
freedom, 98, 108, 111
fruits, 42
funding, 40, 76, 94, 99, 109, 110
funds, 60, 106

G

GAO, 47, 48, 49, 52, 53, 54, 58, 60, 61
GDP, 2, 5, 11, 16, 26, 35, 36, 75, 76, 77, 78, 79, 80, 81, 82, 83, 84, 85, 86, 87, 88, 89, 91, 93, 94, 102, 103, 105
Germany, 31, 91, 92
GHG, 4, 7, 9, 10, 21, 22, 26, 33, 35
glacial ice sheets, vii, 3
global climate system, vii, 3
global consequences, 7
global economic welfare, 82
global economy, 37, 40
global leaders, 111
global warming, vii, 3, 7, 76, 77, 78, 79, 80, 81, 82, 83, 86, 87, 88, 89, 90, 94, 95, 98, 100, 101, 111
goods and services, 9
governance, 66
governments, 41, 49, 57, 68
grants, 41
grazing, 55, 111
greenhouse, vii, 1, 2, 3, 21, 40, 44, 50, 98, 99, 100, 101, 102, 107, 109, 110
greenhouse gas(GHG), vii, 1, 2, 3, 40, 44, 50, 98, 99, 100, 101, 102, 107, 110
greenhouse gas emissions, 2, 40, 98, 99, 100, 101, 102, 107
greenhouse gases, vii, 1, 2, 50
gross domestic product, 2, 57
groundwater, 43
growth, 13, 44, 65, 76, 90, 94
growth rate, 90
guidance, 56, 106
Gulf Coast, 54
Gulf of Mexico, 54, 57

H

habitat(s), 9, 22, 55, 65
Hawaii, 52
health, 3, 4, 10, 11, 24, 108, 110, 112
heat waves, vii, 1, 3, 41, 64
heavy downpours, vii, 1, 47, 50, 64
heterogeneity, 19
history, 20
homeowners, 24, 37, 44
homes, 65, 107, 108
House, 85
household sector, 103
housing, 52
human, 21, 22, 35, 45, 54, 60, 68, 71, 100, 112
human experience, 21
human health, 35, 71
humanitarian intervention, 111
Hurricane Katrina, 64
hurricanes, 40, 44

I

IAM, 25
ideal, 103, 105
IEA, 95
images, 95
imagination, 113
impact assessment, 95
improvements, 8
income, 78, 102, 105
income tax, 102
incumbents, 110
India, 82, 84, 85, 101
individuals, 3, 24, 36, 43
industrialized countries, 16, 34
industries, 10

industry, 13, 43, 44, 45, 57
infestations, 55
infrastructure, 6, 35, 48, 49, 50, 51, 52, 53, 54, 56, 57, 59, 66, 68, 70, 71
initiation, 14
injury, iv
insecurity, 108
institutions, 55
intelligence, 65
International Energy Agency, 90, 95
inventions, 8
investment(s), 3, 6, 8, 16, 26, 35, 44, 45, 56, 92, 101, 102, 103, 106, 108, 109, 110, 111
investors, 39, 40
islands, 66
isolation, 65
issues, 61, 63

J

Japan, 31, 84
jurisdiction, 99

K

Kyoto Protocol, 89, 93

L

land tenure, 108
lead, 12, 43, 53, 72, 86, 103, 105, 106, 111
leadership, 6, 39
leakage, 13
learning, 8
levees, 110
lifetime, 4
light, 88, 99
livestock, 42
loan guarantees, 106
local community, 68
local government, 41, 48, 49, 50, 55, 57, 58, 107
logistics, 70

Louisiana, 57, 58, 61
low risk, 22
lying, 65, 66

M

magnitude, 7, 60, 76, 94, 98
majority, 13
mammal, 22
man, 67
management, 41, 49, 51, 55, 56, 57, 59, 61, 101, 110
man-made disasters, 67
manpower, 68, 99
market failure, viii, 1, 3
Mars, 39
mass, 9, 65, 70, 93
matter, 80, 87, 100, 101, 110
meat, 40
media, 113
median, 15, 36
melt(s), 7, 50
melting, vii, 3, 7, 22
meta-analysis, 14, 16, 33, 34, 36, 37
metaphor, 93
methodology, 51
middle class, 65
migration, 66
military, 48, 52, 53, 64, 68, 69, 70, 72, 99, 111
mission(s), 53, 54, 56, 69, 109
models, 9, 11, 12, 13, 14, 15, 17, 19, 22, 25, 33, 34, 35, 37, 77, 78, 81, 82, 85, 98, 101, 110
moral hazard, 108, 109
mortality, 10
mortality rate, 10
multiple regression, 18
multiple regression analysis, 18
multiplier, 63

N

National Academy of Sciences, 50, 60, 95

National Aeronautics and Space Administration, 49, 52
national interests, 111
National Research Council, 7, 22, 30, 35, 50, 67, 73
national security, 63, 64, 68, 71, 73, 99, 100, 111
natural gas, 102, 103
natural resources, 55
negative effects, 22, 42
negative externality, vii, 1, 3
nerve, 54
Nile, 66
nitrous oxide, 22
NOAA, 58
Norway, 67
NRC, 7, 20, 22, 35, 50, 51, 60

O

Obama, 107
Obama Administration, 107
oceans, vii, 3, 4, 7, 22, 35, 50
OECD, 95
officials, 49, 53, 58, 59
oil, 57, 67, 71, 90, 102
oil spill, 67
operating costs, 105
operations, 42, 48, 55, 56, 57, 67, 69
opportunities, 44, 45, 67
optimization, 35
oxidation, 67
oxygen, 22

P

Pacific, 66, 68
participants, 13, 85
permafrost, 7, 21, 22
permission, iv
Philippines, 69
policy, vii, 2, 3, 4, 6, 7, 8, 9, 11, 12, 13, 14, 21, 24, 25, 26, 33, 36, 44, 56, 65, 75, 76, 81, 82, 83, 84, 85, 86, 87, 88, 89, 90, 93, 94, 95, 96, 97, 98, 100, 101, 103, 104, 105, 106, 108
policy choice, 4, 100
policymakers, 3
political instability, 10
political system, 100, 108
politics, 110
pollution, 4, 26, 45
population, 11, 65, 108, 111
portfolio, 48, 52, 103
positive feedback, 22
poverty, 100, 108, 111
poverty reduction, 111
power generation, 102
power plants, 4, 13, 35, 44, 71
precipitation, 53, 58, 65, 69
preparation, iv
preparedness, 44, 59
present value, 13, 15, 16, 43, 103
preservation, 9
president, v, viii, 28, 39, 48, 56, 58, 59, 60, 71, 98, 100, 101, 112
principles, 107
private investment, 100, 108, 109
private sector, 8, 35, 59, 70, 71, 99, 107, 108
private sector investment, 8
probability, 10, 21, 22, 26, 35, 36, 37, 64, 69, 105
probability distribution, 21
producers, 42
profitability, 40
project, 33, 35, 57, 105
property rights, 108
prosperity, 71, 96
protection, 6, 41, 108, 110
public financing, 8
public goods, 107, 108
public health, 59, 108, 110
public interest, 102
public investment, 98, 99, 108, 109, 110
public policy, 108

Q

quantification, 101, 110

R

rainfall, 42, 50
RDD, 105
real property, 54
real terms, 91
realism, 53
reality, 9, 26, 71, 76
reasoning, 80
recession, 76, 94
recommendations, 25, 48, 56, 59
reconstruction, 109
recovery, 40, 42, 48, 50
recreation, 55
refugees, 66
regional problem, 73
regression, 16, 17, 18, 19, 20, 36
regression analysis, 16
regression line, 17, 18
regression method, 36
regulations, 43, 104
renewable energy, 40, 44, 45
renewable fuel, 103
requirements, 70, 104, 108
reserves, 45, 67, 69
resilience, 48, 49, 50, 56, 57, 58, 59, 60, 61, 70, 73, 100, 107
resource management, 56
resources, 45, 49, 55, 66, 67, 68, 70, 93, 102
response, 11, 22, 46, 60, 67, 69, 111
restrictions, 69
revenue, 102, 104, 105, 106
rewards, 8
rights, iv
risk management, 8, 101, 110
risk profile, 43
Royal Society, 50, 60
Russia, 33, 65, 67, 84, 101

S

safety, 68, 71, 93
salmon, 68
savings, 45
science, 4, 6, 10, 21, 58, 59
scientific knowledge, 7
scope, 9, 51, 88
scripts, 95
sea level, vii, 1, 3, 7, 11, 41, 49, 50, 54, 55, 57, 65, 66, 68, 69, 71, 78
Secretary of Defense, 73, 97
Secretary of Homeland Security, 73
Secretary of the Treasury, 71
security, 64, 65, 68, 70
security threats, 64
Senate, 39, 47, 63, 75, 85, 97
sensitivity, 7, 17, 21, 25, 35, 78
services, 11, 57, 59
shorelines, 107
shortage, 66
showing, 12, 25, 44, 45
signals, 6, 35
significance level, 20
signs, vii, 1
simulation(s), 12, 15, 16, 17, 18, 25, 34, 36
social costs, 35
Social Security, 43
society, 24, 37, 111
soil erosion, 42
solidarity, 109
solubility, 22
solution, 6, 8, 36
South Korea, 91
species, vii, 3, 7, 10, 21, 22, 35, 55
specifications, 19, 36
spending, 41, 45, 48, 49, 57, 60, 98, 99, 102, 110
stability, 100
standard error, 19
standard of living, 109
state(s), 7, 22, 40, 41, 42, 43, 44, 45, 48, 49, 50, 51, 55, 57, 58, 59, 61, 65, 68, 69, 72, 100, 102, 103, 107, 112
state regulators, 43

stock, 22
storage, 13, 71
storms, vii, 3, 41, 64, 68, 70, 71, 101, 109
stress, 24, 66, 69, 70, 71
stressors, 22, 112
structure, 99
submarines, 67
subsidy, 105
substitutes, 51
Sudan, 109
suppliers, 69
suppression, 41
surplus, 108
survival, 55
sustainability, 39
Syria, 66

T

tanks, 71
target, 2, 4, 5, 6, 9, 10, 11, 12, 13, 15, 16, 18, 19, 20, 33, 34, 35, 36, 80, 81
tax base, 103
taxation, 112
taxes, 79, 81, 83, 102, 106
taxpayers, 39, 40, 41, 42, 44
teams, 33
technical assistance, 48, 49, 51, 55, 59
technological advances, 8
technological change, 9, 35
technological progress, 8, 12, 14, 35
technologies, 4, 8, 13, 14, 35, 92
technology, 3, 6, 12, 13, 14, 19, 35, 36, 105, 106
temperature, vii, 1, 2, 3, 4, 5, 7, 9, 10, 11, 12, 21, 22, 25, 26, 35, 36, 37, 58, 67, 72, 78, 100, 101, 106, 110
tensions, 65, 66
theft, 6
threats, vii, 3, 7, 66, 68, 99, 100, 111
tides, 57
Title I, 35
Title IV, 35
total costs, 9, 12, 75, 76, 94
total revenue, 79

trade, 8, 9, 85, 86
training, 52, 53, 69, 70
transformation, 60
transmission, 71, 72
transportation, 49, 50, 57, 70, 71, 103
transportation infrastructure, 71
transshipment, 67
treatment, 9, 24, 34, 57

U

U.S. economy, 39, 98, 99, 102
UK, 91, 95
uniform, 103
uninsured, 57
United Kingdom, 50
United States, vii, 1, 6, 8, 10, 16, 32, 34, 47, 49, 50, 52, 53, 55, 57, 60, 61, 64, 65, 71, 73, 101, 108, 109, 111
up-front costs, 48, 50
urban, 65, 66
urban areas, 65, 66
urbanization, 65

V

vacuum, 93
valuation, 25
variables, 11, 16, 18, 19, 20
variations, 17, 34, 35
Vice President, vi, 97
violence, 100, 108, 109
vulnerability, 40, 42, 54, 70, 109, 111, 112

W

wages, 103
war, 111
Washington, 30, 41, 43, 60, 61, 73, 113
waste, 60, 67
wastewater, 57
water, 10, 22, 35, 39, 41, 42, 49, 55, 57, 65, 66, 67, 71
water resources, 35

water shortages, 42
water supplies, 22
wealth, 111, 112
weapons, 52, 53
weather patterns, 10, 65
welfare, 8
wells, 43
wildfire, 41, 42, 69
wildland, 49, 55, 61
wildlife, 55, 111
wildlife conservation, 111
worldwide, 69

Y

Yale University, 30, 31
yield, 2

Z

Zimbabwe, 111
Zulu, 111